ADVANCES IN BIOETHICS

Series Editors: Edward Bittar and Rem Edwards

Volume 1: Violence, Neglect and the Elderly edited by L. Cebik, G. C. Graber and F. H. Marsh
Volume 2: New Essays on Abortion and Bioethics edited by R. B. Edwards
Volume 3: Values, Ethics and Alcoholism edited by W. N. Shelton and R. B. Edwards
Volume 4: Critical Reflections Medical Ethics edited by M. Evans
Volume 5: Bioethics of Medical Education edited by R. B. Edwards

ADVANCES IN BIOETHICS VOLUME 6

POSTMODERN MALPRACTICE: A MEDICAL CASE STUDY IN THE CULTURE WAR

BY

COLLEEN D. CLEMENTS

Adjunct Clinical Associate Professor of Psychiatry,
Department of Psychiatry,
University of Rochester School of Medicine, USA

2001

JAI
An imprint of Elsevier Science

Amsterdam – London – New York – Oxford – Paris – Shannon – Tokyo

ELSEVIER SCIENCE Ltd
The Boulevard, Langford Lane
Kidlington, Oxford OX5 1GB, UK

First edition 2001

Library of Congress Cataloging in Publication Data
A catalog record from the Library of Congress has been applied for.

British Library Cataloguing in Publication Data
A catalogue record from the British Library has been applied for.

ISBN: 0-7623-0757-9

♾ The paper used in this publication meets the requirements of ANSI/NISO Z39.48-1992 (Permanence of Paper).
Printed in The Netherlands.

DEDICATION

For Helel ben-Shahar

CONTENTS

INTRODUCTION: THE TROJAN HORSE OF BIOETHICS *xi*

PART I LAYING OUT THE ARGUMENT AGAINST THE TOTAL SOCIETY IN A MEDICAL ETHICS CASE STUDY

Chapter 1 What Really Assassinated Hippocrates *3*

The Real Goal to Take Control of the Medical Care System Away from Physicians *3*
The Code Attack on "Paternalism" to Delegitimize Medical Science/Practice Values *7*
False Appeal to Patient Liberty and Empowerment, the "Autonomy" Trick *9*
Legalizing and Regulating Medical Ethics into "Bioethics" *11*
Public Manipulation through Redefinition, Rewording, Trial Balloons *12*
The Liberals "Battle of the Angels" against the "Forces of Darkness" *15*
The Politicizing of the Medical System: Power to Political Ideology *16*
Individual Patients and the Political Total Society *20*
The Repeating Pattern in Educational Institutions, Religious Institutions, the Media, Cultural Institutions, Legal Institutions, the Sciences, the Military *22*
The Total Society *27*

Chapter 2 Bioethics in Historical and Philosophical Context *31*

Existing Histories of the Bioethics Change *31*
From Shaman to Hippocrates *34*
The Hippocratic and Roman (Catholic) Development *36*

Moving through the Discontinuity of 1960–1970 *38*
The Bioethics of the 1970s: Autonomy *40*
The Shift to Utility in the 1980s–1990s *43*
Bureaucratized, Managed Postmodern Bioethics *47*

Chapter 3 The Postmodern Environment of Bioethics *55*

The Conceptual Tripod of Postmodernism: Psychoanalysis, Existentialism/Phenomenology, and Neomarxism *55*
Implications for Reality, Probable Truth, Human Worth, Human Freedom *62*
Re-Lighting the Enlightenment *67*

PART II SPECIFIC ISSUES IN BIOETHICS: ANALYSIS AND VALIDATION OF THE ARGUMENT

Chapter 4 Human Reproductive Medicine: Freedom or Regulation *77*

The Basic Human Values of Fertility and Reproduction *77*
Multiple Births and the Bioethics Response *79*
Old Women and Newborns *84*
Cloning, Sexually Reproducing, and Reading the Human Genome *91*

Chapter 5 Infectious Diseases: Ethics, Experts and Policy in AIDS, vCJD, Ebola Virus, West Nile Virus *101*

The Potential Risks of a Public Policy Mistake *101*
AIDS/HIV as a Politicized Disease in Policy and Bioethics *110*
Doomsday and Ebola Virus *112*
The Policy Response to West Nile Virus *114*
The Ethics of Public Health Policy *120*

Chapter 6 Abortion: Human Worth, the End of Ethics and Political Entitlement *129*

How Political Ideology Has Replaced Ethics in the Abortion Question *129*
How Modern Law Fails Modern Society *135*

Human Worth and Membership in the Ethical Community of Man *139*
When Women Make the Ideologically-Incorrect Choice *146*
The Difficult Question of Personal Identity and Membership in the Ethical Community *149*
The Other End of the Ethical Community Continuum *152*

Chapter 7 Politicizing Brain Death, Treatment Refusal, Physician-Assisted Suicide, Terminal Sedation, Treatment Rationing *155*

Creating the Culture of Death *155*
Certifying Brain Death *157*
The Next Incremental Step: Refusal of Medical Treatment *163*
Kevorkian, Quill and Physician-Assisted Suicide *168*
The New Bioethics Strategy: Terminal Sedation *174*
The Hidden Agenda: Treatment Rationing *177*

Chapter 8 Health Belief Model and Health Delivery Systems *181*

The Beginning of Taking Control of the Purse Strings *181*
The Health Belief Model and the Change of Name to Health Care *187*
The Patient as a Business Unit *191*
The Lack of Statistics and Studies *194*
Should Horror Stories Count *196*
Hope and Utility *199*

Chapter 9 Alternative Medicine, New Age, Classic Shaman Systems *201*

A Shaman Great-Grandmother and Scientific Medicine *201*
Implications of the Burden of Proof *206*
Mistakes on Both Sides *211*
Return to the Age of Herbals 217

Chapter 10 Behavioral Myth, Psychiatric Abuse, and Social Manipulation *221*

The Political Use/Abuse of Psychiatry/Psychology *221*
The New Myths about Unhealthy Human Behavior *227*
Social Manipulation and the Society of Group Therapy *233*

PART III 'BY THEIR FRUITS YE SHALL KNOW THEM'

Chapter 11 Are Patients Better Off than They Were before the Sixties: A Second Opinion and Analysis of Bioethics *245*

Has This Analysis Been too Extreme or Popularized? *245*
The Pragmatic Criterion: Are patients better than they were before Postmodern Bioethics? *251*
The Second Opinion *254*

Chapter 12 Resisting the Total Society, and a New Model for Ethics *261*

Refuting Thrasymachus, Allying with Helel ben-Shahar *261*
Reclaiming the Full Humanity and Breadth of Ethics and Bioethics *271*

INTRODUCTION: THE TROJAN HORSE OF BIOETHICS

In its brief history, bioethics has been looked at critically from an internal, hermeneutic perspective, the critical analysis of the sacred text of bioethics. It has not, however, been critically analyzed from an external perspective, from any other philosophical theory than the power consensus that represents current bioethics – the postmodern ethics of process. Given that postmodernism represents the predominant philosophic school not only in the academies but in the culture, need bioethics be anything other than an internally coherent expression of that postmodernist theory of ethics? My book attempts to show that indeed an external analysis apart from postmodern assumptions is required, and that, as Socrates realized when arguing with Thrasymachus, something more important than medicine is at stake here: "nothing less than the rule of human life."

Bioethics is only a specific case example of a much broader philosophic and ethical problem for contemporary human beings. My focus on case ethics in bioethics has led me to see that bioethics itself is a case example of a general confrontation between Thrasymachus and Socrates on the theory of all values, ethical, cognitive, and aesthetic. What has occurred in bioethics since the 1970s is a refutation of the possibility of all values, except as power-driven consensus-building. Within my naturalistic theory of ethics, what has happened represents the denial of the primary human values of the genetic learning program and the genetic program of emotional responses that constitute our humanity.

I once had a philosophic argument with the Danish philosopher, Justus Hartnack. Professor Hartnack was an internationally-known Kant scholar, and he was also a considerate and very civil man. After his retirement back to Denmark, he sent me a sweet note congratulating me on my first publication in philosophy, the only person to have ever done that. So although we disagreed, I was impressed with his humanity. The argument occurred between three of us, on the question of philosophy of science. Professor Hartnack took the extreme position that one did not have to know science in order to do philosophy of science. The language analyst and I disagreed, and I think we had the stronger argument.

Bioethics, however, in practice sides with Professor Hartnack. Although medicine appears to be the subject, the content of bioethics; medical science and medical values are irrelevant to bioethics in a fundamental way. Bioethics is political, not scientific or clinical. It is not necessary to know anything about medicine in an important way in order to do bioethics. The universal ethical principles of bioethics are not derived from medicine or the biosciences in any way. They come from social conventions, from political positions, from legal strategies, and from a self-selected and power-enforced consensus among bioethicists. Bioethicists themselves admit there is no rational base for their consensus, no valid content to ground it. Their argument is as old as Thrasymachus, that there can be no rational, valid content to ethics. All that can be done is to build consensus, but consensus cannot be achieved through an appeal to reason, actually, because they have already ruled out any rational content. The appeal is not to reason; it is to the use of power strategies not dependent on physical violence but on pressure violence or exclusion violence. Bioethicists would like to think that their consensus follows the political process of democracy, of freedom, but it is not that either. The consensus is not achieved through philosophic education or through allowing all views an equal share of presentation or through freedom to think in new directions and come to new conclusions. The agenda for the negotiating of the consensus is pre-set, public relations replaces education, and opposing views are not brought to the table or are sent from the table quickly. The appeal is not to freedom of thought, expression, and action; but to the acceptance of an established consensus that arose from a self-selection process.

In postmodern bioethics, then, we have lost important intellectual values that developed world culture since that famous confrontation between Thrasymachus and Socrates.

We have also lost important emotional values that have made us fully human since classical humanism. An unusual result of postmodern bioethics is that the medical system has been dehumanized. This will strike some as contrary to the goals of bioethics, which in its early phase stressed the humanities in medicine, with the assumption that humanities professors were better equipped to bring human values into a medicine that was lacking them. Art and literature were brought to the medical schools, poetry was read, therapeutic feelings were elicited. How could I say that bioethics took human feelings out of the medical system?

The most recent proof of this is the strange story out of Britain concerning the guidelines that recommended not using anaesthesia on brain-dead organ donors when their organs were surgically removed. Those guidelines finally (and I should type that in capitals or bold or somehow indicate feeling about it) triggered a confrontation, a non-consensus, with the anaesthesiologists who had before remained silent. They had been silent in order not to jeopardize the public's willingness to donate organs, not wanting to contribute to the shortage of donor organs. They had been silent for years. When they finally spoke, they told of the trauma to the operating room teams (nurses, surgeons, physicians) who watched as the scalpel cut into the body and the blood pressure and heart rate soared in the donor. They watched as the donor body wriggled and moved so violently on the table that the surgeon had to request a paralyzing drug in order to do the surgical removal of the organs. They watched human pain, and they were traumatized – but not enough to speak out until the guidelines appeared recommending no anaesthesia need be given.

There will be quibbling about this. The anaesthesiologists will be accused of self-interest in making money by performing more procedures. The health care system will argue that anaesthetizing a brain-dead patient is a waste of scarce resources, and that the donor has no social utility left to justify such an expense. Neurologists will point out that they are not sure what the pain reflexes or responses mean, and that there may be no perception in an unconscious donor. However, they will also point out that no one actually knows how pain is processed in such patients, and since the patients will no longer be alive, they cannot be asked about their pain perception. The focus might shift to working with the teams to reduce the trauma they feel, rather than considering the brain-dead donor as someone we have an ethical responsibility toward. There are many ways to try and handle this problem, and the most obvious one will likely be ignored.

How can a health care professional watch, acknowledge their internal feelings of great discomfort, focus on their experience – and not consider the experience of the donor? How can they do that for years within the medical system? Why have only a few in bioethics even mentioned this phenomenon, and why hasn't bioethics called for its stop, which could easily be achieved by anaesthetizing all donors? Feelings of empathy, compassion, respect, recognition, love – are curiously absent, yet they once characterized the practice of medicine. Dr. Michel Nostradamus deliberately went to areas hard hit by the Plague, set up houses to treat his patients, managed to do so successfully in some cases, and avoided the disease himself. He left his faculty position and his studies to travel Europe and help. The Knights Hospitallers in Jerusalem set up hospitals for lepers and treated them despite the danger and the social isolation. Between battles in the crusades, they gave love and respect, and their concern was not for their trauma, but for their patients. Human emotions had always been a part of the medical system, not perfect or saintly, but there.

Classic medicine was deeply involved in humanism and the strong emotions that go with sexual reproduction, birth and generation, growth, well-functioning, disease, dying, and hopes for immortality. The fullness of human thought and feeling were reflected in medical practice, and given primary status. That is no longer true, no matter how often current bioethicists talk about "agonizing" over health-care decisions, or "tragic" choices in allocating scarce resources, or "respecting" the patient's autonomy to refuse treatment. There is no personal agony, tragedy, or respect in postmodern bioethics for the human condition. Rather, the framework is one of contract, informational lectures, cost-efficiency within the system, and depersonalization of units within that complex system. Primal human feelings have been replaced by therapeutic concepts reflecting only the consensus segment of the full range of human emotions. The focus, in fact, has shifted thanks to therapeutic self-centering to the caregivers' feelings and traumas, to the wounded physician. Patient and family cannot be treated with human concern in the organizational systems we are creating.

Consider the handling of the brain-dead donor issue. How, in all humanity, could this event have been passively accepted in medicine and bioethics? Nevertheless, bioethics has moved medicine to just such a rejection of the inherent worth of humans, and to the routinizing of the wonders of the natural experience of medicine. Restricting human nature to political praxis, the postmodern consensus of bioethics allows only process and the belief feelings of ideology, or at most, the emotions of power-seeking. This book will demonstrate how humanism was severely weakened in medicine, even while its classic phrases were meaninglessly repeated as justice, respect, personhood, equality, human rights, truth, autonomy, beneficence, and community.

I have been given the opportunity to learn and explore with many who would not come to the conclusion this book argues. I fondly remember my "Dutch Uncle," Joseph Fletcher, and his wife Forrest, who watched the crows outside my living room window. Eric Cassell helped the start of my career and might have understood my concerns about the loss of humanism. Richard Doherty brought me to the genetics division. Roger Sider invited me to psychiatry. Richard Ciccone kept me in forensics. John Romano shared his end-of-life decisions with me and his death affected me even more than I thought it would. Van Potter helped my career in many ways, and I'm sure would hesitate to accept my conclusion, but I'm also sure would agree on a return to updated naturalistic ethics. Ed Wilson's sociobiology and his creativity were a shining model and I hope he will enjoy this work. Judith Landau works closely with me and would find more value in political ideology than I can see. Doris Teichler-Zallen, who wrote and organized conferences with me, would likely also prefer political solutions to my classic solutions. The editors of The Medical Post, who run my columns monthly, have shown by example the value of free thought and non-consensus, and produce one of the very best medical newsjournals. Let me add all the faculty at the University of Rochester School of Medicine and at Rochester Institute of Technology who have been valued colleagues. For all the consensus bioethicists who disagree with me, also gratitude because it is true that one can learn more from those who disagree and are valued enemies than from those who always agree and are valued friends. Special acknowledgement goes to Edward Bittar who has made this work possible.

I should also thank Derek Freeman who fought the good fight against his work on Margaret Mead being professionally censored, and to Hiram Caton who documents such things. Morton Rapp involved me in journalism. Marvin Farber and Tad Clements involved me in philosophy. My family, seven children and eleven grandchildren, are a Greek academy of philosophy all on their own. My sister, Joan Crow, teaches rather than indoctrinates, and knows the difference.

In the end, however, curiosity is the prime motivator. I once wondered what medicine and bioethics were all about. The answer is not always what we want.

Colleen D. Clements, PhD
Adjunct Clinical Associate Professor of Psychiatry
University of Rochester School of Medicine
Adjunct Professor of Philosophy
Rochester Institute of Technology
Columnist
The Medical Post

PART I

LAYING OUT THE ARGUMENT AGAINST THE TOTAL SOCIETY IN A MEDICAL ETHICS CASE STUDY

1. WHAT REALLY ASSASSINATED HIPPOCRATES

THE REAL GOAL TO TAKE CONTROL OF THE MEDICAL CARE SYSTEM AWAY FROM PHYSICIANS

What is in a name? In postmodern bioethics, a great deal. For the first years, bioethics, biomedical ethics, or medical ethics never clearly stated its actual goal. It was a forming interdiscipline having difficulties selecting a name to describe itself adequately, as Van Rensselaer Potter nicely documents in his book, *Global Bioethics* (1). It turned out that there was a good deal in a name, since Potter's initial suggestion of bioethics implied an actual interdiscipline where the biological sciences and value theories would conjointly develop a new field based on the probable truth they both contained. The alternative two names both placed "ethics" as the major noun or class, with "biomedical" or "medical" as its qualifier, proclaiming a conceptual mistake of modern ethical theory that asserts that scientific knowledge cannot be foundational to an ethical theory (2). The preferred use of either of those terms implied that the field of ethics had primacy over medicine or the biosciences in a discipline that claimed an important role for ethical analysis. Now that the foundational primacy of ethics has been established to the satisfaction of practitioners of this new field, some are more comfortable again using the term first suggested by Potter, "bioethics," but they do not mean the term in the way that Potter intended (personal correspondence).

There has also been an interesting process in who should be considered part of the consensus in bioethics or biomedical ethics. Initially, a number of contrasting ethical theories were represented by the various professionals entering the new field. There already was a large body of writing on Roman Catholic medical ethics which needed to be considered by the new discipline, but for the most part, that development of ethical theory was quietly disregarded or seen in opposition to a "secular" medical ethics. Classic objective theories of ethics such as Thomistic or pagan Natural Law were eventually characterized as failing to meet the needs of a multicultural, spiritually diverse, democratic society in the United States, and failing to meet the requirements of a global biomedical ethics. The philosophic reverence for the worth of a human being, as exemplified in Kantian and Neo-Kantian ethics, was reduced to a notion of "respect" that Kant would have rejected as meaningless within his system. Respect became another code word for allowing patients, as political and legal entities, to have any choice they made honored and facilitated. Although Paul Ramsey, a Protestant and Neo-Kantian theologian (3), was prominent in early medical ethics, his concept of respect was applied only when it fit the emerging consensus, and could be used to place certain classes of people identified as victims of oppression or discrimination or helplessness into a restricted category, or as an unqualified tool to enhance patients' rights. Pragmatic or situation ethics was also initially included with Joseph Fletcher's involvement in medical ethics (4). But Fletcher too was eventually removed from relevance by a benign neglect of his position, perhaps because it placed agape (general love) for the individual over formal or bureaucratic ethics. Situation ethics can be used as the antithesis of institutional or procedural ethics, and was also not compatible with the developing consensus on biomedical ethics. I will examine this essentially process or political ethics consensus in more detail further in this chapter. For now, I only wish to maintain that there was a consensus developing which extruded ethical outliers, and to suggest that this development was not accidental, but controlled by a specific goal and a specific position in ethics which was far from multicultural. There was premature closure on bioethics.

Within the last decade, the specific goal of this new discipline has finally been articulated by members of its consensus, and I wish to make that goal as apparent as I can. First, one of my co-authors, Doris Teichler-Zallen, wrote an article outlining part of the new goal, to prepare the public to accept certain policy decisions or practices in the biological sciences that the public would initially not be prepared to accept (5). This is not an investigative goal, an attempt to analyse ethically in a neutral stance in regard to both conclusion and application. It is a public-policy formation goal, generally with the policy already proposed by experts in the field and now to be communicated to the public in a way that will gain the public's acceptance. It is fraught with potential dangers, the greatest of which is an abuse of the scientific and philosophic process. It can easily become advocacy, in the model of political advocacy. If the purpose of bioethics is political policy formation, we would need to look very critically at the process and have standards that achieved the high requirement of the classic scientific method or the classic philosophic rules of reason. We would have to avoid premature consensus, appeal to authority, coercion through bureaucratic or committee process, emotive use of language and other constraints that are not apparent in the field of bioethics. For example, we would want to avoid analyzing the case of a brain-dead, pregnant woman by accusing her physicians of treating her as an "incubator only," or the too-easy conflation of "persistent vegetative state" with "vegetable" when analyzing medical futility.

For another straightforward statement of the goal of bioethics, David Rothman's historical study, *Strangers at the Bedside*, is a careful account (6). Rothman describes how clergy formed the Society for Health and Human Values and then secularized the group to the Committee on Health and Human Values, whose purpose was to use the humanities to introduce current social values, generally neoleft values, into medicine. He traces the very political development of bioethics from the 1960s and describes Senator Ted Kennedy's hearings on quality of health care and issues of human experimentation. Rothman finds the goal of such hearings to be one of underlining the need for outside decision makers in medicine, composed of "the center and the left." He concludes that the worthwhile goal of bioethics was from the start to take control of decision-making away from physicians and bioscientists and to place it in the hands of lawyers, professors of humanities, politicians, and social scientists. His book details the step-by-step orchestration of that goal, from dialysis committees (God committees, hence 'playing God'), the Senators Kefauver/Javits hearings on drug regulation, the FDA and NIH/USPHS rules on consent and research review leading to the institutional review boards that were fueled by Dr. Henry Beecher's polemic in *The New England Journal of Medicine*, The Daedalus Conference with participants skewed to an emerging consensus, the Harvard Brain Death Committee fueled by expert conclusions in *The Journal of the American Medical Association*, The Joint Commission on Accreditation of Hospitals" patient bill of rights conjointly with the political National Welfare Rights Organization, Senators Mondale/Ribicoff bill for a commission of science and social values, the activities of the Kennedy Foundation, Senators Kennedy/Mondale move for a national commission on medical ethics, the Kennedy Institute's active legal involvement in the Karen Ann Quinlan case from which arose a long list of statutes on end-of-life, and the documentation of trial-balloon cases to generate such legal responses. Absent from his long historical analysis are any activities by conservative or religious groups with varying conclusions, or any alternative ethical points of view. Rothman's book has the advantage of frankly stating the agenda of bioethics: to wrest control of medical decisions from physicians and place them with the favored group, policy activists in law, politics, and the humanities.

At first, H. Tristram Engelhardt, Jr. and Kevin Wm. Wildes (7) seem to suggest there may not be a secular consensus in bioethics, but they do admit that there was an attempt to replace the plurality of religious bioethics with a particular secular bioethics. Still, they felt secular bioethics also fell victim to a plurality, finding there is now more secular bioethics than major religions. I doubt that is the case, because the premature consensus for a particular secular bioethics still seems operative, one based not only on a secular theory of ethics but on a morally pluralistic setting, implying that objective secular ethics is as impossible as objective religious ethics. That they finally fall back on a procedural morality, or process ethics, is not surprising, although their dismissal of the possibility of an objective, probable theory of ethics should be logically surprising. But there is some sensitivity in their approach as they discuss why current bioethics experts seem successful in their application of a secular bioethics. They concede that underlying this success is in fact a shared, particular secular "moral vision." They even name this an ideological understanding, perhaps without fully realizing how political that ideology is. Both believe in the legitimacy of informed consent, the free market, and limited democracy based on the permission of the participants. Both feel permission alone is enough and that a content-full ethics is not required. What would be required is a "peaceable" consensus, the faith in a peaceable community (8) that may be more irrational than any of the competing content-full ethical theories, and that seems very much relative to the time and place of post-seventies U.S. society.

On the other hand, Daniel Wikler is even more political and more forthright in his description of the goal of bioethics (9). He uses the term "health policy workers" to describe the activists with whom bioethicists must interact and within which some bioethicists actually belong. He sees bioethics as an "activity," a mission to reform medicine to make it more responses to patients' interests and preferences (two very different and often contradictory things if we think about it, I would argue). He admits that partisan attachment to certain positions on bioethics issues was a criterion of expertise in bioethics, and my own experience at my medical school would attest, sadly, to that. But Wikler is not overly dismayed by this cultural relativism in ethics. In the end, he embraces it for health policy as does Michael Walzer (10) for social criticism, accepting the attempt to make a moral critique founded on values intrinsic to our culture, history and tradition. None of this gives us any claim to truth in ethics, so it will finally reduce to an ethical nihilism, as does Engelhardt and Wildes, Rothman, and Teichler-Zallen. Bioethicists, however, do not appear to be bothered by a final nihilism concerning human values. In Chapter 3, I will look into some of the reasons for this casual attitude about the end of ethics.

For now, I want to summarize the goal, which is actually a power play. The goal is to take decision-making power, hence most power, away from the professionals in medicine and biosciences and give it to professionals who share a common political ideology from the humanities, law and politics. The goal, really, is to politicize from a particular political position the institutions of medicine and bioscience. In place of the criticized religious ethics that cannot serve for a diverse multiculture, we now have a particular political position which bioethicists somehow believe can serve for the diverse multiculture, even though many of the elements of that multiculture may be in stark disagreement with that particular political ideology. The logic of this escapes me, but the political power strategy of it does not escape me. It greatly concerns me.

THE CODE ATTACK ON "PATERNALISM" TO DELEGITIMIZE MEDICAL SCIENCE/PRACTICE VALUES

From the beginning, bioethics made extensive use of the term "paternalism" in negatively analyzing physicians' behavior toward patients. The word was one of the many dyslogistic terms used in the field, carrying much emotional baggage. It, along with "patriarchal," reflected a feminist political agenda to devalue males; but it also was its own a priori argument against the standard Hippocratic code of medical ethics.

Robert Veatch made this use of the term quite clear and explicit in an appendix to his book (11). Paternalism was the code for an attack on the objective value system found in medicine since the time of Hippocrates, a system whose main value was to achieve a good outcome for the patient. Good outcomes can be evaluated with some probable truth value and do not depend on patient perceptions, wishes or choices alone. If such a professional value system was accepted, then the experts in medicine and bioscience would be relevant and powerful in determining which decisions were ethical and which were unethical. The professional, the physician, could make important and valid statements about what the best interest or need of the patient might be. The professional, the scientist, could say something highly significant about a risk assessment or cost/benefit analysis, and not have that be culturally relative, a product of consensus social values. The professional could define and control his or her profession, something arguably crucial to the existence of a profession (12).

This was not acceptable within the framework of the assumed bioethics ideology. That ideology first had the goal of replacing the professional physician or scientist with the professional lawyer, politician or humanities professor and/or activist. Secondly, that ideology denied the possibility of a content-full bioethics in order to render useless any power base for dissenting positions. The only power base allowed was to be a consensus among selected bioethics experts who would use negotiation and a particular political process to ground ethical conclusions and policy formation. All competing ethical systems, including the Hippocratic system, would have to be made irrelevant, either by change and cooption or by attack on coercive, paternalistic, father-knows-best, idiosyncratic physician practices and on war and postwar abuses not restrained by scientists and their scientific method. As Rothman points out, this attack was orchestrated with public relations expertise and highly successful with the media. He gives Beecher's use of the press and of a professional journal as an example, although he is not critical of this public relations component of bioethics (13). Beecher's article was rejected by *The Journal of the American Medical Association* since it was not a careful research article but more in the genre of an editorial or opinion piece. It has been legitimately criticized on its scientific merits and fairness to the experiments discussed. However, *The New England Journal of Medicine* did publish his paper, and Beecher called the press to alert them to the significance of the article as he saw it.

Based on the Beecher article, a cascade of political events occurred, resulting in changes in obtaining consent (from informal talk between physician/researcher and patient/subject to formal informed consent forms) on both the national funded-research level and in hospitals. Paternalistic professionals could not be trusted to do the right thing or work in the best interest of the patient. Decisions had to be taken from their hands for two reasons. One, they could not be trusted to be ethical and had demonstrated according to the new bioethicists that they routinely did evil things. Two, even if they were ethical, they had no bases to make value judgments other than their own idiosyncratic opinions about right and wrong. There was no professional or philosophic system of ethics that had validity. There could be no content to ethics.

But decisions had to be made. Professional ethicists might maintain that their bureaucratic system could make such decisions, based on political ideology about process and common cultural values, but this might be seen as replacing one set of biased professionals with another set. It would also be a risky power strategy with a heavy burden of frontline accountability.

FALSE APPEAL TO PATIENT LIBERTY AND EMPOWERMENT, THE "AUTONOMY" TRICK

The value of "free" choice is an important value in most ethical theories. It has a balanced but central role in natural law theory and naturalistic ethics, a primary role in existentialist and phenomenological ethics, in Neo-Kantian ethics, and seemingly in postmodern ethics. In a political context, it reflects the social values of a democratic form of government, given Thomas Jefferson's warning that it can only be a desired form of government if the population is an educated one, not an uneducated or manipulated one. From its beginning, the field of bioethics appealed to free choice, using the term "autonomy," but in a way that Kant would not have accepted. Patient autonomy was set above most of the forming consensus of values that would make up the Kennedy Institute's Principles of Medical Ethics (14). The "good" was defined not by consequences or by intentions, not by objective needs and interests, not even by traditional cultural values. The "good" was any choice a patient made freely, without coercion, after being informed. This definition did require that the individual be competent to make choices, and not a member of the special class of individuals who lacked capacity to consent. There were many candidates for the definition of competence, but over time it became a very minimal standard of capacity, approaching the forensic definition of competent-to-stand- trial. Within hospital or clinic settings, or even on the streets in the case of individuals with mental diseases or ordinary citizens who refused to be taken to the hospital, most individuals were seen as meeting a minimal standard for capacity to consent. (The only exceptions were therapy patients consenting to have a relationship with their physicians, or, as it later developed, patients choosing medically futile treatment.)

It would seem, then, that patients were given liberty and empowerment (another political and ideological term) by the new bioethics, but this would be deceptive. The field of bioethics carefully and only implicitly restricted such liberty. Patients were given a libertarian freedom to refuse treatment. They were not given the same freedom to request and receive treatment, especially treatment that was ideologically disapproved. Fertility or reproductive medicine treatment is an area where bioethics is every bit as paternal as it was claimed medicine was. Almost all the advances in treatment have been met with strong criticism by bioethicists who call for increased regulation and even moratoria on research. Individuals seeking such treatment are not considered sufficiently competent to make such important decisions, although they would certainly be competent and free to choose to die by deciding to have supportive treatment withdrawn.

Individuals deciding they wished to have medical treatment even though they were terminal would also find their freedom to choose indirectly restricted on a routine basis (negative messages about resuscitation, requests to fill out a living will form, difficulty with admission to intensive care units, restricted information on treatments available). In many circumstances, they would even encounter treatment refusal based on medical futility as determined by the physicians and hospitals.

There are even restrictions in place in office practice. Regulations and changing standards of practice make obtaining antibiotics beyond the patient's choice. In outbreaks of infectious diseases, it may be difficult to access either antibiotic prophylactics or vaccines, in my experience even when there has been contact with an index case or a sibling of the index case. Tests are now carefully rationed, even if the patient wishes to have a test and even if a test might have been previously indicated.

Hospital discharge is also not within the patient's choice. In fact, patients have been discharged by use of an eviction notice served by a sheriff, in one case on a Gulf War veteran.

So we are dealing with a logically inconsistent application of an ethical principle, which generally suggests that the principle has an ambiguous meaning. Only some ideologically-acceptable autonomy is given by bioethics, for purposes other than the expression of the individual's free will or moral agency. The care patients can ask for is now more restricted and regulated than it was in the past. The power base for an individual patient is now less than it was. By emphasizing in ethics and law the autonomy of the lone individual patient, both family support and power and expert advocacy have been reduced or eliminated. The patient has no ally within the current health care system who can support his or her needs and interests. Nor can an ally protect the sick individual who is laboring under the handicap of illness from being manipulated except with much difficulty. The family cannot easily intervene. The bureaucratic system, which the patient does not know or understand, models the behavior of those who are part of it, and outside resources have been for the most part excluded. Instead, the individual must rely on other complex bureaucracies like committees for accreditation, quality reviews, state health departments, policies and procedures.

The individual may also have to rely on an ethics consult or an ethics committee, and on occasion, the ethics committee meets on a case without the patient's awareness. Often a conclusion is included in the chart, and those meetings can influence how patient care is done. What has been reached in the end is the replacement of physician-knows-best with either professional ethicist-knows-best, or regulatory systems influenced by bioethics-know-best. In either case, autonomy (except to refuse treatment) was not what it seemed, and without accountability/liability the bioethicist has replaced the medical professional as decision maker. The goal to place other professionals in control of medicine and bioscience has been achieved in a circuitous way, and the patient may be less free than was the case forty years ago. This, in fact, fits nicely into Engelhardt's call for a bureaucratic ethics (15). In a bureaucracy, autonomy is not the operative word.

LEGALIZING AND REGULATING MEDICAL ETHICS INTO "BIOETHICS"

The Hippocratic system of medical ethics was based on the interaction between physician, patient, and often family. The new bioethics is not. From the beginning, the political aspect was dominant, and political power is achieved through statute or legal precedent in a court decision. The easily-anticipated result was a decrease in attention to philosophic or religious ethics and an increase in appeal to the legal and regulatory systems. Rothman's book nicely describes all the various court cases and agency regulations that mark the history of current bioethics. He describes the critical period of change, as he calls it, in medical ethics from 1966–1976. The three major goals are outlined: (1) collective decision making, (2) formality in decision making, and (3) power to professions outside medicine and bioscience to frame the normative ethical principles of medicine and bioscience (16). I've discussed the last goal and sketched how it has been achieved through the new bioethics. The first and second goals convert ethics into law and politics. Collective decision making is a characteristic of a political ideology which reduces the individual to a unit in a particular power class, making the single individual irrelevant to social processes, and requiring membership in a class in order to achieve some action or recognition. These collective classes are then weighed in terms of their power base and/or skill at public manipulation. Can these classes define the terms and therefore control the argument and determine the conclusion? Can these classes push through legislation or through hearings influence setting up task forces and committees? Can these classes determine who will be selected for membership on such committees or agencies? Can these classes access the political power levers?

Bioethics moved easily into political access and power as Wikler and Rothman demonstrate. It may now be coopted by health policy experts, although it has the advantage of the usefulness of the strategy of "taking the moral highground." The strategy of taking the moral highground is not competent philosophic or religious ethics, however, and that does (and should) disturb Wikler. The risk of politicizing ethics is well-expressed by two classic examples: Epicurus who advocated retreating from the Greco-Roman world into his garden of ethical contemplation, and Marcus Aurelius fully committed as emperor and philosopher fighting to maintain Greco-Roman (and therefore world) knowledge. My sympathies are with Marcus Aurelius, but there are risks in that choice, and current bioethics illustrates them. The bioethicist can become an apologist for enshrining the ideology of the moment into government systems that are static, not a valuable or pragmatic move. Bioethics can become abused for power strategies, something we quickly criticized the U.S.S.R.'s psychiatrists for when applied to dissidents. But if psychiatry can be abused, so can bioethics, something I will show in Part II.

As for defining ethics as law, that danger is extremely high for any society. If ethics is reduced to law, we lose the ethical base for distinguishing between good and bad laws, and for justifying civil change or disobedience (17). The law also has its own characteristics that make it not very suitable for fulfilling the role of ethics (18). It is too powerful in its consequences (for example, loss of liberty) to apply it to every decision or act we wish to disapprove. It can create a totalitarian environment. It is difficult to change and shares the problem of static lack of adaptability with bureaucratic political systems. It can justify the most awful national mores.

Nevertheless, law and government are the remedies almost immediately turned to by the current bioethics. In discussion of the issues in reproductive medicine, bioethicists have suggested repeatedly in the media that this medical specialty needs both more regulation and new legislation to resolve those issues. The question of physician-assisted suicide worked its way through the courts as an issue of constitutional privacy, until the Supreme Court wisely decided there was no such privacy right that it wished to create.

PUBLIC MANIPULATION THROUGH REDEFINITION, REWORDING, TRIAL BALLOONS

There is a larger ethical issue than bioethics, the value of a free mind under assault by public relations techniques that are very sophisticated and by a media that can be used very effectively to transmit them. Free thinking is a critical component of ethics, one of the high-ranking values that is a precondition for being human, for knowing good and evil, and for doing the right thing. This value, one of the values most important to me personally, has always been at stake in bioethics, and I want to outline the public relations technique that is a model for this.

One of our forensic psychiatry fellows had a seminar with me in which we discussed bioethics. For him, the issue of public manipulation was the most important concern. He had worked for systems that made significant use of such manipulation, and he knew the advances made in the psychological sciences that made it such a danger. He was a Kurd minority in Iran before he fled the country, and while he considered the dangers in Iranian society, he felt the danger in the developed countries of thought control was greater. It was an interesting cultural perspective, and I began to look more closely at a pattern that can be seen in most bioethics issues. I began to consider redefinition, rewording and trial balloons in a new context apart from fallacies of inductive reasoning.

In Part II, I will analyze Ebola virus more fully, but here I want to sketch the steps in making Ebola virus an American household word. I chose this example because it is more neutral than other bioethics issues, but it is easily applied to the other issues as well. Activists concerned with brain-death legislation, living wills, physician-assisted suicide, reproductive medicine, treatment refusal, involuntary commitment, plutonium experiments, and AIDS/HIV have followed the same basic pattern.

1. An academic conference was sponsored by Rockefeller University and NIH on emerging, resurging and resistant viruses. In bioethics, conferences that were primarily academic were sponsored by CIBA on ethics and transplantation, by Daedalus on ethical issues in medicine and research, a symposium was held highlighting the Johns Hopkins Down's syndrome infant, the Ross Conference focused on ethics in newborn nurseries to name only a few. I was a member of the organizing committee for the Nazareth Conference on medical ethics myself.

2. An academic book was published by Stephen Morse, part of the committee that set up the conference (19). I was co-author of an academic book from the Nazareth Conference (20). Professional credibility had now been established for the bioethics issue, though objectivity is not guaranteed or even achieved in this process. Conference attendees can be selected for a particular position to be advanced, risk assessment may not be well done, and an activist position can permeate the entire process without being corrected by the scientific method or rules of reason. But this is only the first and not very effective step, because neither such conferences nor books will reach the general public in sufficient number to initiate public calls for action. Only professional respectability has been accomplished, but that is an important requisite, even if only a skewed consensus.

3. A trial balloon or media case was selected and promoted in the media. The Ebola case was the Reston, Virginia outbreak among lab monkeys, in a laboratory in this suburb of Washington, DC. Richard Preston was given access as a science reporter, and did a story on the Reston outbreak that emphasized the great risk and alien nature of Ebola virus. Other trial balloon cases have

become famous in bioethics: the Beecher list of unethical scientific experiments on subjects who were not given informed consent forms; the Karen Ann Quinlan case that allowed life support treatment withdrawal in a persistant vegetative state patient, the Nancy Cruzan case that allowed withdrawal of nutrition and hydration in a PVS case; the Janet Adkins case of an Alzheimer patient requesting physician assistance with her suicide; the "Debbie" case in *JAMA* that supposedly involved a direct lethal injection by a physician; the "Diane" case in *NEJM* on the issue of physician-assisted suicide; the Baby Doe case on refusal to treat or feed a Down's syndrome newborn, etc. Some of the trial balloons will fly (Quinlan, for example), others will not (the anonymous "Debbie" case). These cases will appear in the popular media and then be commented on in professional journals, not as research articles but as editorials or opinions bathed in the same aura of expertise as research articles. Some of these cases will also move into the legal arena.

4. Continuing popular press coverage and discussion of these cases was the next step. In the Reston/Ebola case, Preston went on to write a best-seller (21) and *60 Minutes* did an hour presentation with Biohazard Level Four suits. The public has now been given "information" on the bioethics issue, although that information may be skewed or even deceptive. The public perceives a high-risk or serious problem, and, if the pattern has functioned as intended, will now support a call for action. Generally, the actions have already been formulated by activist groups who are waiting to implement them, so the public will be given a restricted menu of options, those favored by the groups that began the process. Input on other options will be constrained.

5. Some response to public pressure generated by the pattern occurred. Congressional hearings were held and more funding was made available to virologists and the Centers for Disease Control and Prevention for research and monitoring of emerging and resurging pathogens. In general, this was not a bad response, but Ebola posed no major risk to justify such policy priorities. It had at this point never caused an epidemic, but rather isolated outbreaks that did not spread, even in a suitable environment for spread like the Congo. In other bioethics issues, the responses can be: congressional hearing with an eye toward federal legislation or funding, state legislative action in the form of regulations or laws, specific court decisions that establish legal precedent, executive orders, revising of professional codes or guidelines, moratoria on research.

Based on this pattern or model, I find the value of free thought at great risk in bioethics. This pattern for identifying issues, for giving "correct" analysis, for moving the public to a preset conclusion and action, is a pattern of strategic and sophisticated manipulation.

THE LIBERALS "BATTLE OF THE ANGELS" AGAINST THE "FORCES OF DARKNESS"

Why has the issue of manipulation (the political strategy is now called spin-doctoring) not been raised as a major concern for bioethics? I think the answer lies partly in the politicizing of ethics, the comfortable premature consensus that Engelhardt denies and then admits as a solution, and the rise of postmodernism in our academic culture. In Chapter 3 I will go into postmodernism in detail, so in this section I only want to address its practical interaction with bioethics.

I cannot think of a better metaphor than one inadvertently presented to one of my co-authors and I a number of years ago. We had been asked to review a book on law, psychiatry and ethics by a rising star and law professor, Michael S. Moore (22). We had disagreed with some of the premises and arguments in the book, but it wasn't until the last chapter that we finally recognized how deep the conflict was and the degree of emotion tied to the other side. In our careful reading of the arguments before reviewing the book, neither of us had expressed our ideas in terms of demonizing those who disagreed, or in the model of an apocalyptic confrontation between the archangel St. Michael and the fallen archangel and demon Lucifer. Such a form of thinking would not have occurred to us. But there in the last chapter of the book was a summation of all who disagreed with the author as being part of the forces of Darkness. Those who shared the author's position were described as being on the side of the angels, and instead of an ethical and rational analysis of the premises of both sides, we were now being exhorted to join the battle on the side of the angels, against the forces of darkness.

Moore's approach first shocked me, but then I realized that bioethics was not about well-constructed philosophic reasoning, or about verifying premises in an argument, or about scientific investigation of problems. Bioethics had become part of the culture war, in which tactics like dehumanizing or demonizing the enemy (those who disagreed with the goal or the consensus) were predictable and routine. Activists always assumed their side was the side of the angels. I had never seen the demonizing and battle tactics so frankly expressed in an academic publication, however, nor had I realized at that point how far from classic academic disciplines bioethics had moved.

Our review was critical of the author's position, though not in the framework of a cosmic battle between good and evil. Still, the review was criticized because it was a negative evaluation of the author's position, though not because our objections did not have merit. Over continuing philosophic analysis of bioethics, I came to realize the strength and prevalence of the activist tactics and the difficulty of the struggle to maintain bias-corrected, valid reasoning. The task of ethics has always been a difficult one, even in the classic theorizing, but philosophers have attempted to approximate the standards, and have always held the standards even when failing to come up to them. There has been a change, though, as postmodern culture turns more to the view of Thrasymachus that talk of ethics is merely the imposition of the biases of the elite and powerful. Although both Thrasymachus and Socrates lamented such a view of ethics, and Socrates attempted to show it was not a logically necessary conclusion, that view is now embraced by postmodernism for reasons I will later dissect. The view means the end of reasonable and viable ethics and value theory, but most bioethicists do embrace it, with Engelhardt one of its champions (23). It is a position about the possibility of content-ethics that flows from, and is very compatible with, the politicizing of social institutions, including academic institutions.

THE POLITICIZING OF THE MEDICAL SYSTEM: POWER TO POLITICAL IDEOLOGY

We usually do not think, or did not think, of medicine and politics as equivalent. However, since the 1960s, the medical care system or institution has been more and more politicized and less and less based on the biosciences and care of patients. That will seem a strange statement, because the beginnings of bioethics focused on respect for patients, criticized the dehumanization of patients and insisted on seeing the patient as person, emphasized patient freedom or autonomy, and talked about quality care for all individuals. It will be necessary to look at what was actually done, and how it was done.

One of the power bases for the political consensus in bioethics was feminism, or rather, the current dominant concept of feminism, since there have historically been many varieties of feminism. The current dominant concept is a highly political one, based on ideas of power and seeing most social and personal interactions as power struggles. There is a heavy bias against the characteristics of masculinity that are either biologically or socially based, a need to view males having those features as oppressors, and an over-generalized, probably false description of most Eurasian societies as patriarchal. The goal is to replace male dominance with female dominance, at least for a select group of professional females. In bioethics, this move to power is cloaked in the form of redressing oppression and sexism on the part of a male-dominated profession.

One of the effective strategies has been to focus on sexual abuse and harassment. Both terms have been kept deliberately vague, so that the abuse or harassment is determined by the perceptions of the individual who feels abused or harassed. In surveys taken by special task forces or professional bodies, the questions are so generally worded that it would be difficult to construct a list of specific behaviors that are ethical and specific behaviors that are unethical (24, 25). No social scientists have been vocal in criticizing the construction and wording of such instruments, and the results of these surveys have been used in journals and courtrooms as scientific evidence.

The first reform centered around sexual affairs between patients (almost always women) and physicians (almost always men). Based on a misreading of Freud's concept of transference-love, reformers argued that due to the strength of the transference between patient and psychiatrist, it was not possible for any patient to consent to having a sexual affair with her psychiatrist, that it was automatically abuse and exploitation of the patient, that the therapy patient must lack any capacity for consent, and that the codes of ethics must incorporate this understanding. If Freud is carefully read in his first description of transference-love, it should be apparent that he did not really believe the patient was so snared by transference that she was incapable of consent, or that the feelings of love the patient reported were always resistance and neurotically-dominated (26). He did, however, suggest that the patient be told that was the case (proposing a medical deception that has continued to the present). Transference now took on the strange character of a coercive force that made any woman a helpless victim should the physician choose to take advantage of her. Psychoanalysts struggled with this ethical issue, generally allowing some latitude for human emotions on the part of patient and physician, and requiring that the analyst terminate with the patient before beginning a love affair. Marrying the patient might have repercussions for professional advancement, but medical licenses were not removed. Still, there was the psychoanalytic view that a patient, given the power of the transference, should always or for some specified timeframe be considered as a patient by the physician, even after termination.

With the emergence of bioethics, that psychoanalytic description of the patient in relation to the physician began to apply to all physicians, not just to psychoanalysts. It has now spread to all specialties in medicine, even involving emergency room physicians in questions about boundary issues (27). "Boundary issues" is a further extension of the meaning of sexual abuse/ harassment into non-sexual behaviors as well. Medical codes of ethics began to reflect this new application, with pressure from professional feminists, special clinics for sexual abuse cases, media cases, malpractice trials, and even trials for "rape by authority" (Quebec, Canada). Both language and behavior could qualify as unethical.

I was involved as a consultant for two cases. In one, the physician had jokingly told a new patient who seemed uncomfortable that she needn't worry, he didn't intend to rape her. In another, two patients charged a psychiatrist with rape because, even though they had wanted the alleged sexual affair, the authority or power of a physician constituted rape. This was a test case that had been selected by feminists who had waited and planned for such a case, in an attempt to extend the interpretation of a Quebec law that allowed charges of "rape by authority." The woman who was the crown prosecutor, worked closely, and had been working in the past closely, with a woman psychologist who had recently published a book based on information given to her by women who had been solicited to contact her with their experience of being sexually abused by physicians. The two women plaintiffs in the case had contacted this expert witness prior to reporting the alleged sexual affairs. In addition, the prosecutor and expert witness worked together during the trial contacting feminist and sexual-abuse groups in the U.S. The judge, however, ruled against the crown, both factually and on the basis that lack of capacity to consent had not been demonstrated.

Many states have changed their laws specifically to disallow any capacity to consent, no matter what the actual psychological state of the patient is. By statute in these states, patients have been deprived of their legal competence to determine what relationship they will have with their physicians. It is necessary, of course, to reduce women patients to helpless victims, to "rag dolls" as they frequently call themselves, to infants without the freedom to decide on their actions. With freedom, an accountability or responsibility is implied that would make it impossible to argue that a consenting adult with no major psychiatric disorder or mental disability has been sexually abused or raped.

Harassment is a broader charge with even less specified meaning. Any verbal statement by a physician could be interpreted by the patient as harassment, so there is a good deal of thought and speech control accompanying the new bioethics.

The feminist political agenda has also resulted in research being broken into sexual components, with women' diseases being described as underfunded. As a result, government funding agencies now give some priority to grants for women's health issues, imposing a political reverse discrimination standard on the scientific selection of research proposals. The problem with this approach is that basic research tends to be non-sexual, and it is from basic research that most promising information about body systems and effective interventions comes. Breast cancer stands a better chance of being cured if we look to the molecular level of all cancers and then apply that sophisticated understanding to treatments for breast cancer as it affects women and prostate cancer as it affects men. It is true that there are differences in how females and males respond to diseases, just as it is true that there are differences in how ethnic groups or races respond to diseases – and more importantly, just as it is true that there are differences in how individuals respond to diseases. Variability within a species is a biological given.

The most extreme version of feminist politics applied to medicine may be the feminist argument against reductive science. This argument (spurious within systems theory in general) was applied to AIDS/HIV and I encountered it in personal communication when I was director of our department's AIDS Training Program/NIMH. It simply would not have occurred to me when we wrote our book on AIDS (28). The argument was that masculine science tended to find the pathogen and to reduce the disease to the molecular level, so that the identification of Human Immune Deficiency Virus as the cause of AIDS was a mistaken masculine science. Feminine science, on the other hand, would look at the whole person, the functioning organism at its family and social level, and see the causes of AIDS as refusal of men to allow their women to use condoms, drug use because of underclass position in society, and other social issues. Masculine science was reductive; feminine science was wholistic.

Having written a fair amount on systems theory, I think I am in a position to reject such a simplified, political, and non-systems approach (29–32). Reduction is a very necessary component of any systems approach, and it is critical to identify the molecular processes of AIDS. A systems approach is not the special turf of feminists, but is an epistemology used by any person in the biosciences or other sciences. To deny the importance of the work of Luc Montagnier in discovering the likely virus causing AIDS is unnecessary and discriminatory. But its worst effect is to undermine the standards of science with the introduction of such sexual discrimination and political bias. Nor should we make the mistake of thinking it would be easy to understand the higher levels of organization that include social practices/interactions, or that intervening at this level would be effective, except perhaps in a totalitarian society.

There are other examples of politicizing the medical care institution. The Principle of Justice, part of the Kennedy Institute's "mantra," is a very political concept, based on late twentieth-century liberalism that is actually a combination of neomarxism and postmodernism. It has been used to influence health policy planning, including insurance proposals such as managed care, rationing of access to tests and technologies, cutbacks in the development or implementation of technologies, and even utilitarian choices of limiting care to the elderly or terminal patient in favor of well-baby clinics (the favorite source of reallocation). Reallocation of resources and the redefinition of quality care as cost efficient care have all been political positions subsumed under this ethical principle.

The ongoing argument on abortion has also been a political rather than ethical argument. It involves legal positions on the privacy right created in the Griswold decision (33), and the latest Supreme Court decision (34) concerns whether state governments can regulate medical procedures, a decision that gives physicians enormous power in one specific area, the terminating of pregnancy and death of the fetus. Whether physicians should have such unrestricted power, the same question asked in physician-assisted suicide, has been bypassed for the political emphasis on a privacy right.

INDIVIDUAL PATIENTS AND THE POLITICAL TOTAL SOCIETY

One of the consequences of politicizing the medical institution, as the process in fact also occurs in other central social institutions, is that politics is a strategy for obtaining and maintaining power based on control of large classes of individuals. Each of us belongs to some political demographic group, to some political ideology, to perhaps some activist organization. It is within the group or class that our political power resides. But the sick individual, the patient in the hospital or office, becomes a much more intimate biological unit than a unit of a class. Bioethics used to criticize calling the patient in A-103 the pneumonia in A-103 because it dehumanized the individual by seeing only his or her pneumonia. It was assumed that this was unacceptable depersonalization. But bioethics has created a far more serious depersonalization of that patient. A-103 is now looked at by a utilization reviewer to see whether length-of-stay has exceeded the general length-of-stay for pneumonia, and if it has, usually unknown to the patient, the physician and hospital administrator will have a discussion about insurance coverage. A-103, on entering the hospital, will be given routinely and bureaucratically a list of patients' rights (which sounds good) and also a living will form with the implication that the hospital would like the form to be filled out. A-103 will be given an informed consent form before any procedures are done. Whether there is good news or bad for A-103, his or her physicians will conduct a minicourse on the disease as if the patient were in a college classroom. A-103 will be evaluated in terms of protocols developed from large classes of patients and will be assumed not to deviate from those protocols. Of course, before A-103 was admitted to the hospital, he or she had to meet the guidelines of the insurer, based not on the actual physical condition of A-103, but on the insurer's willingness to admit for certain statistical classes. Even A-103's pneumonia is no longer that individual. It was selected as the best diagnosis for insurance purposes and does not differentiate between individuals having multiple disease processes. And if the prognosis is poor, A-103 may find that he or she will not receive hoped-for treatment because the statistical success rate makes it "medically futile." After forty years of bioethics, it may have been better to be the pneumonia in A-103.

Can A-103 function self-protectively in this new health care system? It will be difficult. First, A-103 is quite ill with pneumonia. He or she may not be lacking capacity to consent, but all energy should be going toward recovering, and the brain does get affected by illness even if the person can meet the minimal legal standard of competence. A-103's family might like to help, but they are restricted for two reasons. Unless they fully understand the new health care system, they will not have the necessary information to make decisions that would benefit their family member. Even if they can make such decisions, the law will not allow them to interfere with their family member's choices. Health care professionals also are restricted in their advocacy. First, they are fearful of being accused of paternalism or violating patient rights, and they have been educated to believe they have no special expertise in the decisions the patient has to make about health care. It would only be their idiosyncratic opinion or the imposition of their value system. Second, the economic pressures within the new health care system model behavior, whether we wish to believe that or not. Good people may perform bad things in a bad system. Third, most physicians have now been exposed to bioethics education or training, and that education tends to be done within the consensus of the bioethics field. It will reflect the political ideology of the consensus, and that ideology supports the politicized health care system. Can the patient or anyone appeal to an ethical standard apart from the bioethics consensus, one that has content? Bioethics has argued no one can. There is no ethical base from which to criticize bioethics, it argues.

The conclusion is that individuals have very little power within the health care system, despite forms on living wills, bills of rights, or informed consent. The health care system has become part of a politicized Total Society, along the lines of the total institutions that sociologists have described. Within total institutions, individuals lose their personal identities and are programmed to act in ways compatible with the institution. It is a form of totalitarian organization that is subtle but effective. One of the arguments of this book is that medicine has become a part of the Total Society and now lacks a separate base of its own with which to criticize some of the activities of that Society. But a separate base was once the definition of a profession, once the definition of social institutions other than the political institution.

THE REPEATING PATTERN IN EDUCATIONAL INSTITUTIONS, RELIGIOUS INSTITUTIONS, THE MEDIA, CULTURAL INSTITUTIONS, LEGAL INSTITUTIONS, THE SCIENCES, THE MILITARY

It was with pleasure that I read a reporter's book on the U.S. military that was making the same point for the military institution that I have been making for the medical care institution. Stephanie Gutmann's *The Kinder, Gentler Military* is an analysis of how the New Army came into existence, and the implications of this politicizing of the military for our society (35). Carol Gilligan argued that Gutmann approved the values of violence which had brought so much damage to the world, and that women in the military would be superior to such a warrior society, but she did not dispute Gutmann's substantive arguments, only her values which were different than Gilligan's and the feminist consensus (36). The same argument could be brought against Gilligan. She approves too easily the value of pacifism.

Gutmann came late to her analysis, as did I, remembering the day in 1995 that a phone call to Army public relations allowed her to see that something strange was happening to the U.S. military, embarrassingly late in the day, as she describes it. She visited boot camps and comes up with the same conclusion I do for medicine: "What stands out starkly in the revised curricula of modern boot camps is ideology." Gutmann documents that training and military readiness are sacrificed to this ideology, and that speech and behavior must conform to it. Her book looks at all branches of the armed services to make the argument.

But perhaps the most important analysis her book contains is her conclusion that most of the change from the Old Military to the New Military closely parallels a general cultural movement in the U.S. She calls the activist groups such as the National Organization of Women the "personal is political crowd," because she sees their goal as moving private life into political discussion and then into courts, judges and regulatory agencies. She thinks it was inevitable that this "personal is political" movement would eventually change the military as well. They had spent most of the 70s and 80s, she argues, focusing on the workplace, the schools, and the home, but finally because of their political consensus determined it was time to also move into the military institution and politicize it.

With constitutional safeguards within the legal system, it was difficult for the political activists to immediately effect a change. The strategy of using sexual abuse or harassment would not be effective as long as there were legal protections. By the middle of the 80s, however, she documents that harassment became discrimination under the Civil Rights Act of 1964, under Title VII, and the definition of sexual harassment had been changed to include any verbal or physical behavior of a sexual nature that creates a "hostile, intimidating, and offensive environment." This strategy was not the same as that used in the medical institution, transference, but it has worked as well. And, in fact, female consent in the military is rejected using similar arguments found in medicine. Power imbalance or male dominance was used to argue away consent.

Gutmann discusses the results of this politicizing of the military by first pointing out the documented evidence that standards have been relaxed to accomodate the difference between the male body and the female, and that some of this relaxation will be dangerous in war zones that can now easily include the rear. Former Marine commandant General Charles C. Krulak has described the "three-block war" in the media in some detail. Although the term "peacekeepers" sounds as if lowering infantry and helicopter standards so that women can participate would not be a problem, that term is dyslogistic and does not describe the actual military maneuvers that General Krulak anticipates for the future.

Second, she talks about the military importance of group morale, the single-sex group bonding that is critical in the history of warfare. Unit cohesion is an expression of that morale on the lowest level of the military machine, which is still a human machine. She calls it "love," the selfless love in which sex is irrelevant and protectiveness is paramount. But she argues that men and women are hardwired to cohere in very different ways, that sexless group bonding very quickly becomes paired bonding which is not compatible with unit cohesiveness. The professional value on which so much military success is based is now threatened because of a political experiment with the military. Even worse, she shows how the top-ranking professionals in that institution gave up one of their primary values and allowed outsiders to construct values for them. She accuses the leaders of this profession of criminalizing the warrior spirit because it was seen as "anachronistic, abusive, insensitive, elitist." One ex-naval officer told her the brass "refused to defend their own culture." In medicine, that same pattern can also be seen. Once the core values are sacrificed, it is a simple procedure to completely politicize the institution. Again, there is no value base on which to stand to criticize what is being done as unethical.

Gutmann was the first writer I had encountered who saw the pattern of politicizing all important social institutions, although she saw it from primarily a feminist political agenda. My argument is that the politicizing movement is broader than just a feminist goal. A sophisticated and organized conceptual framework called postmodernism is the impetus for this politicizing, and it extends its changes through all the institutions Gutmann named, plus others. She was not aware of the extent of political change in the medical care system or some religious insitutions, nor the law, sciences, media and cultural institutions.

The sciences have only partly resisted postmodern politicizing. There have been special conferences on the correctly-identified attack by postmodernism on science (37), and a properly-applied method of scientific reasoning can maintain the value base of that profession to an extent. But the process has occurred, particularly in the biosciences and medical sciences, especially in ecology. Poorly designed research, if its conclusions fit the political views of the society, will be promoted. It is fairly easy to list what views can influence the acceptance or rejection of research:

1. research that shows human beings are the primary cause for a particular environmental degradation,
2. research showing some success for natural as opposed to artificial treatment of diseases,
3. research showing the negative impact of population, consumption, and pollution,
4. research showing that less expensive medical care is better or equal to expensive medical care,
5. research showing there is questionable genetic determination of behavior,
6. research showing that poverty or social discrimination is the most important factor in health,
7. research showing some positive effect for spiritualist practices,
8. research showing an egalitarian human nature,
9. research showing negative impacts of technology,
10. research proposing an apocalyptic vision, such as global warming, diversity depletion.

Gutmann mentions schools as having been politicized, and this is an accurate evaluation, both for primary and higher education. Academic institutions in the 60s and 70s were among the first institutions to become politicized into a particular political ideology. It was suggested that American universities adopt the model of European politically-active universities, and the suggestion was readily accepted. The value system of academia that should have stressed imparting of world knowledge, unbiased analysis, and freedom to think and pursue any hypothesis was abandoned for limitations that were described as progressive, liberal and holding the moral highground. Systems were put in place to assure that both professors and students did not deviate too far from the political consensus of those reforming/changing the institution. Again, the rational base, the valuing of reason, was replaced by political and social causes popular at the time. More important, underneath the causes was an ideology of postmodernism that devalued the Greco-Roman bearer of world knowledge/culture, the European bearer, and the Asian bearer (with the exception of Buddhist renunciation). This general politicizing of universities made it easier to change the medical schools affiliated with the universities. Bioethics as taught in medical schools reflects the postmodern ideology of the university, not the reasoning of philosophic schools.

Primary and secondary education has also now been politicized, with states mandating certain political values courses and training, and less time left for traditional or classic subjects. Those mandated courses are described as teaching cultural diversity or world culture, but a look at the curriculum indicates they are actually an attack on world knowledge, world culture, and a promotion of one of the major postmodern values: all conversations are equal, there is no standard to evaluate discussions (38). This is not quite true, because European conversations are evaluated, and negatively. By the time the products of this training reach undergraduate university, as my experiences with my Rochester Institute of Technology students illustrates, those who have European backgrounds talk with loathing about European evils and have no pride or esteem for their backgrounds. They are apologetic. The primary and secondary school goals of re-enforcing self-esteem only apply to some groups and not to others. This also is a political tenet of postmodernism.

The consequences within the educational institution are that careful and critical thinking values are removed as a base for the institution, and that correct knowledge of the past experiences of human beings operating in the world is made unavailable for support and comparison. Again, the base for criticizing political ideas, provided by such educational institutions, has been made unavailable. The politicizing of academic institutions, rather than leading to freedom of thought and liberty, leads with high probability to indoctrination instead.

We might suppose that religious institutions would continue to perform their historic task of supply a supernatural base for evaluating and criticizing political power. But the situation is clouded even here, because many religious institutions have been themselves politicized. The traditional main-line churches joined the political activism of the 60s and 70s and replaced creedal beliefs with the social gospel and social activism. Perhaps reflecting the modern problem of belief in a supernatural level, little attention is given to defending the traditional views on the nature of God, heaven and hell, sin and soul. Attention has moved to a loving, ambiguous deity who watches approvingly as members do charitable deeds or effect social change. Those religious institutions that do work within a doctrinal level have been marginalized by society as "religious fanatics," rightwing Christians trying to impose their beliefs which they foolishly believe are more true than others. Since postmodernism has no place for mysticism or metaphysics, these institutions are seen as at best irrelevant and more commonly as enemies of autonomy. Their discussions and conversations are not treated as equal. Sometimes, in fact, they are banned in public settings. Because such religious institutions have resisted being politicized or have instead the wrong politics, they are not part of the forming consensus and may be irrationally vilified. The danger of an opposing religious base is recognized within this political consensus as a high danger.

Religious institutions have been allowed to play a role in bioethics, but there has been a selection for the social gospel forms, rather than those which have not yet lost control of their value system to political ideology. In our medical school, we have entertained Seneca shamans, conferences on chi-gong, New Age spiritualism, and liberal theologians who wish to include "spirituality" as part of medical care. "Spirituality" sounds as if it might be an independent base against political ideology, but it would be a mistake to think that. The term is so vague that it is nearly meaningless. It describes belief, but that belief need not be supernatural or even of a higher power. It focuses on meaning and community, but in a content-less way. Its practitioners must be flexible and open to any meaning and community, and like bioethics, it remains content-less.

Legal institutions have difficulty maintaining a separate base or value system apart from the political views of the time, but it is nevertheless one of their primary functions (or used to be). The classic Aristotelian philosophy of law gave way in the twentieth century to first a logical positivism about the law (39), and then to a proposed utilitarian philosophy of law (40). From positivism and an economic utilitariansim quite different than the British Utilitarians proposed, it is a short step to postmodernism. The value system of the law has been undercut and replaced with a positivist efficiency that became a utilitarian cost efficiency with strong political undertones. The battle for the U.S. Supreme Court may in the end be a battle over content-less and process ethics, or a content-full professional value system. This battle is important for medicine because bioethics has reduced so many value decision to legal decisions.

The media and cultural institutions were the vanguard of postmodernism and remain its strongest institutional base outside the political system. Ideology can and has effectively used both institutions to its advantage. By defining its goal as health, public safety or other agreeable good things, government and its agencies have been allowed access to public manipulation that is unprecedented. Some messages are identified as "public service," but others are embedded in situation comedies, plays, movies, art exhibits. Public relations and its possibilities are recognized by all activist groups, but certain political ideologies have more access than others, and those proclaiming the moral highground, like bioethics, have easier access. Bioethics (the consensus bioethics) is very much a media event. The media, however, in its reporting capacity, has never done a thorough analytic coverage of bioethics itself, or described the consensus formation and its alternatives.

I have very briefly indicated the degree of politicizing of major social institutions, of which the medical institution is only one example. I wish to place this discussion of bioethics in such a context, however, because I do not think it is possible to understand either the history or philosophy of bioethics apart from this context. Much of bioethics needs to be decoded.

THE TOTAL SOCIETY

What I have sketched in this chapter has sometimes been labelled by my colleagues as itself a postmodern analysis. That is half true. One of the concepts of postmodernism is that all human intellectual history has political overtones, and that politics must be considered as an important factor. In other words, postmodernists do politicize everything, even their own disciplines and work.

But I am not a postmodernist, because teasing out the political factors is only the first part of my task. Having identified the political factors, I do not then become an ideologue and either defend those I approve or form strategies for attacking those I dispprove. My base for evaluation is not political and in my philosophy everything does not reduce to the political component. If I had a content-less ethics, that might be true, since it does follow that if there is no content, everything becomes a meaningless, valueless process of some sort. That process can be a committee negotiation or democratic vote, but there is no reason for it to be limited to that, no reason at all within the system of postmodernism. We could just as easily choose a process of armed confrontation with an Uzi as a committee negotiation, and in fact both are coercive, but in different ways. Those who control committee member selection and committee agenda are just as coercive as those who use armed force, and perhaps less honest.

No, I am not limited to such a nihilism in bioethics or ethics or science or reason, one that has to fall back on political ideology and processes for gaining power. I do not think the postmodern argument against content has been successfully made. I do have a value base from which to criticize the consequences of totally politicizing social institutions. Such a Total Society is unethical and contrary to human nature. The value of free thought rather than ideological thought has always been a prerequisite for ethics, although our myths have vacillated on that. The myth that Eve was weak and sinful for picking the fruit of the Tree of Knowledge of Good and Evil is not a myth compatible with human nature. Even the myth that Lucifer refused to bow down, thought for himself, had pride and wished to be treated as a free equal, and therefore deserved being thrown down forever is not compatible with human nature.

My content-full analysis of bioethics comes to the conclusion that its role in creating a Total Society is an abuse of ethics, an attempt to reduce ethics to consensus politics that must be intellectually challenged. I think I have been in the field long enough to take up that challenge.

REFERENCES

1. Potter, V. R. (1988). *Global Bioethics: Building on the Leopold Legacy.* East Lansing, MI: Michigan State University Press.
2. Sincer, P. (1981). *The Expanding Circle: Ethics and Sociobiology.* New York: Farrar, Strauss and Giroux.
3. Ramsey, P. (1970). *The Patient as Person.* New Haven, CT: Yale University Press.
4. Fletcher, J. (1966). *Situation Ethics: The New Morality.* Westminster, Philadelphia.
5. Teichler-Zallen, D. (1989). Regulating research: A tale of two technologies. *Teaching in Society.* Vol. 77: 377–386.
6. Rothman, D. J. (1991). *Strangers at the Bedside: A History of How Law and Ethics Transformed Medical Decision Making.* New York: Basic Books.
7. Engelhardt, Jr., H. T., & Wildes, K. W. (1999). In the beginning: The emergence of secular bioethics. In: R. B. Edwards & E. E. Bittar (Eds), *Advances in Bioethics* Stamford, CT: JAI Press.
8. MacIver, R. M. (1986). The deep beauty of the Golden Rule. In: E. D. Klemke, A. D. Kline & R. Hollinger (Eds), *Philosophy: The Basic Issues* (2nd ed.) (pp. 407–413). New York: St. Martin's Press.
9. Wikler, D. (1994). Bioethics in health policy: What methodology? In: E. Bittar & N. Bittar (Eds), *Principles of Medical Biology,* Vol. 1A (pp. 23–36). Stamford, CT: JAI Press.
10. Walzer, M. (1987). *Interpretation and Social Criticism.* Cambridge, MA: Harvard University Press.
11. Veatch, R. M. (1981). *A Theory of Medical Ethics.* Cambridge, MA: Harvard University Press.

12. Freidson, E. (1970). *Profession of Medicine*. New York: Dodd, Mead; and (1972). *Professional Dominance*. New York: Atherton Press.
13. Rothman, D. J. (1991). *Strangers at the Bedside: A History of How Law and Ethics Transformed Medical Decision Making*. New York: Basic Books.
14. Beauchamp, T. L., & Childress, J. W. (1994). *Principles of Biomedical Ethics* (4th ed.). New York: Oxford University Press.
15. Engelhardt, Jr., H. T. (1986). *The Foundations of Bioethics*. New York: Oxford University Press.
16. Rothman, D. J. (1991). *Strangers at the Bedside: A History of How Law and Ethics Transformed Medical Decison Making*. New York: Basic Books.
17. Feinberg, J., & Gross, H. (Eds) (1995). *Philosophy of Law*, Fifth Edition. New York: Wadsworth Publishing.
18. Feinberg, J., & Gross, H. (Eds) (1995). *Philosophy of Law, Part Two: Liberty, Part Three: Justice, Part Four: Responsibility*. New York: Wadsworth Publishing.
19. Morse, S. (1993). *Emerging Viruses*. New York: Oxford University Press.
20. Teichler-Zallen, D., & Clements, C. D. (Eds) (1982). *Science and Morality: New Directions in Bioethics*. Lexington, MD: Lexington/Heath.
21. Preston, R. (1994). *The Hot Zone*. New York: Anchor/Doubleday.
22. Moore, M. S. (1984). *Psychiatry and Law: Rethinking the Relationship*. New York: Cambridge University Press.
23. Engelhardt, Jr., H. T. (1986). *The Foundations of Bioethics*. New York: Oxford University Press.
24. Gabbard, G. (Ed.) (1989). *Exploitation in Professional Relationships*. Washington, D.C.: American Psychiatric Press.
25. Perr, I. D. (1989). Medicolegal aspects of professional sexual exploitation. In: *Exploitation in Professional Relationships* (pp. 211–27). Washington, D.C.: American Psychiatric Press.
26. Clements, C. D. (1987). The Transference: What's love got to do with it. *Psychiatric Annals*, *17*(8), 556–60.
27. Gabbard, G. (Ed.) (1989). *Exploitation in Professional Relationships*. Washington, D.C.: American Psychiatric Press.
28. Landau-Stanton, J., & Clements, C. D. (1993). *AIDS, Health and Mental Health: A Primary Sourcebook*. New York: Brunner/Mazel.
29. Clements, C. D. (1983). The lesion and the function: Setting up the reductionist problem. *Perspectives in Biology and Medicine*, *26*(3), 433–40.
30. Clements, C. D. (1976). Stasis: The unnatural value. *Ethics*, *86*(2), 136–44.
31. Clements, C. D. (1999). Systems bioethics: The future of epidemics and medical ethics. In: R. B. Edwards & E. E. Bittar (Eds), *Advances in Bioethics* (pp. 31–60). Stamford, CT: JAI Press.
32. Landau-Stanton, J., & Clements, C. D. (1993). *AIDS, Health and Mental Health: A Primary Sourcebook*. New York: Brunner/Mazel.
33. Griswold v. Connecticut, 381 U.S. 479 (1965).
34. Stenberg v. Carhart, No. 99–830 (2000).
35. Gutmann, S. (2000). *The Kinder, Gentler Military: Can America's Gender-Neutral Fighting Force Still Win Wars?* New York: Scribner.
36. Gilligan, C. (5/17/2000). *Make War, Not Nice!* New York Times.com/books.
37. Browne, M. W. (6/4/1995). *Scientists deplore flight from reason*. New York Times.
38. Oakeshott, M. (1991). *Rationalism in Politics and Other Essays*. Indianapolis: Liberty Press.

39. Hart, H. L. A. (1995). Positivism and the separation of law and morals. In: J. Feinberg & H. Gross (Eds), *Philosophy of Law*, Fifth Edition (pp. 56–72). New York: Wadsworth Publishing.
40. Posner, R. (1995). The concept of corrective justice in recent theories of tort law. In: J. Feinberg & H. Gross (Eds), *Philosophy of Law*, Fifth Edition (pp. 374–84). New York: Wadsworth Publishing.

2. BIOETHICS IN HISTORICAL AND PHILOSOPHICAL CONTEXT

EXISTING HISTORIES OF THE BIOETHICS CHANGE

There is a need to reconnect the philosophic standards of the Hippocratic ethical tradition with modern evolutionary biology and general systems theory. Faced with the nihilism about reason, ethics and nature found in current bioethics, its abdication to power in the form of politics, it is necessary to return to the roots of medical ethics and trace, from a conceptual point of view, how medical ethics was transformed in bioethics. The need for an ethical method that corrects for such political, economic, and social biases should become clearer and need not be seen as a hopeless task.

Postmodernism tends to either ignore history or to rewrite it for political purposes, so we should not be surprised that there are few critical histories of medical ethics/bioethics. This chapter will look at those histories that are available and will trace a conceptual framework for looking at the transformation of medical ethics into current bioethics.

There are three histories of medical ethics worth considering: (1) Kelly's *The Emergence of Roman Catholic Medical Ethics in North America*, 1979 (1); (2) Rothmans' *Strangers at the Bedside*, 1991 (2); and (3) Macklin's *Enemies of Patients: How Doctors are Losing Their Power and Patients are Losing Their Rights*, 1993 (3). Engelhardt has also taken a look at the beginnings of current bioethics with an analysis of the philosophic positions that led to this process ethics (4). Each is a different attempt to understand the discontinuity of the 1960s and 1970s that produced our current bioethics, but all are oriented to the twentieth century and do not sketch a full historic context. I intend to go back to the beginning, perhaps because my great-grandmother was an herb doctor and possibly one with shamanic roots, but more importantly, because the ethical history can teach us something to help in evaluating this discontinuity.

David Kelly has the most detailed and historically-rooted work, tracing the changes in Roman Catholic medical ethics in North America in a very scholarly way. His scope, however, is the very specific tradition of Roman Catholic religious ethics with its extensive literature on medical ethics. He wishes to understand the break that changed, he argues, this Roman Catholic medical ethics from an objective ethics informed by reason and faith to a subjective personalist ethics focused on psychological concepts. It might seem unfair to his position to characterize the new personalist ethics as a subjective one, since I believe he would consider it an addition to natural law concepts. But in essence, it is subjective. For example, in the case of twins where one twin required a kidney transplant, the ethical decision now should include how the healthy twin would feel if he were not able to save his brother from death. That interior feeling should rank as high as objective concerns about damaging or mutilating one's body. The issue of internal and external consequences is important, since it can produce a utilitarian theory of ethics that is cold and deadly when it excludes internal consequences. The same criticism could apply to natural law theories of ethics if the focus does not include the internal. But there is more going on here, because casuistry has always included internal motivations, and the ethical defense of a just war states the critical importance of the reason or motivation for the war. The shift to personalist ethics has to mean more, and it does, which is why I describe it as subjective. In fact, the shift downgrades the objective natural law conclusions and elevates the interiorizing, psychological conclusions to primacy. When that occurs, individual feelings and motivations take precedence in reaching an ethical decision, and this leads to a subjectivism in ethics. That personal subjectivism was quite compatible with the cultural change that was occurring in society, and at least for North American Roman Catholic medical ethics, Kelly has succeeded in documenting the gradual shift to postmodern bioethics.

Ruth Macklin was an early expert in current bioethics. Her book (3) describes current economic and political forces in the health care system, and the problem of coercion that such forces represent. She remains concerned with individual patient autonomy and understands that managed care systems or cost-cutting agendas could subtly pressure or force patients into withdrawing treatment or refusing treatment. Her focus is on the present managed care insuring plans and how this economic environment threatens to change the health care system put in place in the 1970s, so there is little indepth history. Her solution is to reinforce the autonomy principle, to continue to apply the ethical principles of the 1970s with more vigor, and her faith is that such stronger application of postmodern bioethics can resist the new forces. She does not, however, see the relevance of looking at the theory of postmodernism to understand what is happening in the health care system, and once dismissed this analysis of theory as academic, rather than helpful to bioethics. Since she chooses not to consider the postmodern base of bioethics, she will fail to see how meaningless the autonomy principle is, and how easily it can be replaced by a cost-cutting justice principle or a medical futility principle. The true fragility of the autonomy principle as a safety factor will not be apparent. The irony is that such bioethics principles are themselves determined or defined by politicoeconomic forces such as Macklin wishes to oppose, rather than being firmly based in objective standards of reason and science. These content-less principles are word-forms into which can be placed any meaning that is convenient or desired for a particular time, or if the meaning is made clear, they are still powerless because based on nothing but politicoeconomic forces.

David Rothmann does a thorough historical description of the rise of current bioethics. He goes back to the war years of the 1940s and then takes the development step-by-step from the 1960s on, focusing primarily on the political and legal aspects, rather than the philosophic ethics aspects. He accepts the conversion of ethics into law, economics, and politics as the only appropriate response to technological and social trends that have led to physicians becoming "strangers at the bedside." He partly approves of the removal of decision-making power from physicians to be placed in the hands of a particular political consensus of politicians, lawyers, and professors of humanities. In fact, he is quite plain about what that political consensus is and calls it a centrist-left combination. For him, the change requires the continuing bureaucratization of bioethics and medicine, which he accepts. Given the need for a secular, multicultural bioethics, he approves of the influence of political forces in bioethics, but assumes those political forces will be a specific political ideology (of the new left). Unlike Engelhardt, he does not see that this secular political ideology is only one of many diverse ideologies. This is critical for bioethics, because it is only an historic accident that the political ideology of bioethics is now neomarxist. In the future, bioethics could reflect any of the ideologies, and within postmodernist theory there is little with which to oppose such a change. If one chooses Thrasymachus, where might (or power) makes right (bioethics), one has no foundation on which to oppose the changing of the guard, the vagaries of power shifts.

It is time to look more deeply into history and philosophy so we can answer questions such as:

> How did physicians become redefined as facilitators, gatekeepers, and civil servants?
> How was the premature closure on bioethics reached?
> How was ethics used as a political strategy and is this an abuse of ethics?

To begin to answer those questions, we need a map of the territory and a sense of a timeline. A lawyer I once did consulting work for showed me how he approached his legal strategies for a case. He constructed a timeline of all events, rolling it across his conference table and mulling over it. I found that tool very useful in expert consultation and testimony, and I think physicians, policy planners, and the public would find it just as useful. Attempting to diagram a timeline is frustrating, and I have made three attempts (5–7). For this chapter, I think it would be better to take the reader through time with words, but some might find the previous diagrams useful.

FROM SHAMAN TO HIPPOCRATES

The shamanic tradition of Eurasia existed until recently in its archaic form among the aboriginal groups in Siberia and was carefully described by Dioszegi in his travels through that area in the late 1950s (8). This form of health care relied both on herbal treatment and some surgical intervention for its botanical-biological component, and on mysticism that explained diseases as some cosmic imbalance. From frozen mummies in the Altai and in the Karim basin of Western China, we also have some evidence of horsehair sutures. We should not assume that health care was entirely mystic or spiritist among the early Scythians of that region. There is also the mention of a medical figure, Huatuo, who apparently goes back to 2000 BC, and the Chinese Great Herbal that dates about 3000 BC. The Greek oracles also conducted herbal and mystic healing.

Egyptian medical care is closer to the Hippocratic physicians, with the figure of Imhotep and the Ebers papyrus of 1500 BC. Careful reading of Egyptian medical texts shows they were quite good at diagnosis, even if we use our modern standards, and their treatments were not that removed from modern medicine (9).

Returning to the shamanic tradition, however, there was a mystical belief system that categorized shamans into Black Shamans and White Shamans, and that was concerned with a mystical order to reality, as fits Claude Levi-Strauss's work (10). Black Shamans were more powerful than White Shamans, but they were also capable of evil if they so chose. White Shamans, while not perceived as wishing to do evil, could be fakes, mentally ill, eccentric. Dioszegi found a pattern of shamans both being respected and feared in their communities, and not being really part of the community. Becoming a shaman was not really a choice for the individual. The calling seemed to run in families, sometimes skipping a generation, but with a strong familial component. Often, the individual was informed through a dream or a sign that he or she was meant to be a shaman, and frequently this was associated with a life-threatening, mysterious illness. If the individual resisted the calling, he or she would eventually go mad or die. Eventually, most accepted their fate and became shamans.

Despite fond memories of my great-grandmother, Mary Pfeiffer, there is a dark side to the shaman. It is difficult to trace her ancestry, because she lied on the church records, and there is even some possibility that she might have been Seneca Indian, though I don't really believe that is true (an expert on Seneca features insists that they are present). This dark side is composed of mystery and fear, and perhaps even quackery. Their communities perceived the shamans as powerful, eccentric, potentially deadly, and untrustworthy. This lack of trust and fear of harm was a reasonable response, since shamans had their own special ethical standards and those standards did not prevent them from using their power as they chose, not necessarily for the best interest of their patient. One took a great risk in going to a shaman, because the shaman could choose to heal or to kill, and there was little protection against this. The Senecas, for example, killed one of their healers because they believed she had killed a man who came to her for healing, and was therefore a witch. The woman, Kauquatau, fled to Canada, but the reservation chiefs lured her back with false promises and then killed her on the Buffalo Creek Reservation in New York as soon as she set foot on reservation land (11). The Allegheny Seneca Reservation had a flood of witchcraft hysteria associated with members of the tribe who might be witches, and it was difficult if not impossible to tell a shaman healer from a witch.

The botanical component of shamanism also carried some risk. The leaves of botanical herbs are not uniform in the active ingredients they contain, the ingredients have unintended side effects even though "natural," and it is possible to kill as well as cure with herbs. Without a scientific reduction to the molecular level, herbalists had to rely on trial-and-error in treating, and the errors could be fatal.

The shamanic tradition, without adequate ethical safeguards, was one of the strong traditions at the time of Hippocratic medicine, Egyptian medicine, and even Roman medicine. It did not disappear after the Hippocratic code of ethics, nor after scientific medicine, but has been resurrected (and misunderstood) in bioethics, where it is supported as "non-traditional," "alternative," "New Age healing," or "spirituality in medicine," and where it has found a home that does not require meeting the standards of the scientific method in the U.S. National Institutes of Health. As an alternative to physician "dominance," it fits the ideology of bioethics and can also be used to erode the power of scientific medicine. It also happens to be considerably less expensive than scientific medicine, which appeals to another goal of current bioethics, cost efficiency and downsizing of the medical care system. In addition, with the support of the biopsychosocial model as misinterpreted to repudiate reductive bioscience (12), and with an emphasis on mystical wholistic medicine and its psychological underpinnings, it nicely fits the postmodern framework of bioethics.

But the dark side of shamanism, something I feel fairly well-placed personally to evaluate but indicated in most good anthropological descriptions, has not been adequately considered.

THE HIPPOCRATIC AND ROMAN (CATHOLIC) DEVELOPMENT

The Hippocratic tradition was an ethical advance from the shamanic tradition and addressed some of the ethical concerns found in that tradition. Hippocratic medical ethics constructed ethical obligations that gave physicians trustworthiness. One of the major ethical obligations of this code was that there be a good outcome for the patient, as Pellegrino stated (13). If it were not possible to achieve a good outcome with medical care, the ethical obligation was then not to leave the patient in worse shape than when he or she first came to the physician, in others words, "at least do no harm." Current bioethics has redefined good outcome to meet statistical outcome protocols and include cost factors, and has shortened "at least do no harm" to "do no harm." The former does not guarantee an ethical outcome for the individual patient, and the latter is a nonsense phrase. Given the structure of reality, the systemic interconnections, no action taken (and also no action not taken) is capable of meeting an absolute standard of not causing any harm. The result of such an ethical principle is an inertia, but we will see that this emphasis on harm or risk in such an extreme way also served the purposes of the political ideology.

Returning to the actual sense of the Hippocratic ethical tradition, a good outcome was something that could be objectively ascertained and did not depend on feelings of satisfaction or free choice. The physician was seen as an expert in medical care, and within that context, as an expert on the good for man. He could be an advocate and in fact was required to be an advocate. He was also part of an ethical brotherhood with a well-developed value base. Hippocratic ethics, in the Aristotelian sense, was objective ethics. This classic ethics produced a number of ethical conclusions, but none of them were seen as universal ethical principles. They were derived from the nature of man and from the environment of the polis or city-state. Killing or poisoning patients was unethical, and providing poison for others to kill was also unethical. The Hippocratic tradition knew the importance of trust in the physician-patient relationship, and understood that like the Black Shamans, a physician who would sometimes kill could not be trusted. It was too great a power to exercise. Doing high-risk abortions was also prohibited because of the likely harm to the woman. A currently very misunderstood ethical rule was that physicians should not enter the households of their patients and take advantage of the situation to have sexual relations with family members or slaves. This was not an absolute prohibition against a love relation developing, or a notion that family members and slaves were not competent to consent. It was a practical prohibition against exploitation. Similarly, the rule of confidentiality required that physicians not gossip in the marketplace about what they had learned in their patients' homes. It too was not absolute in the sense of a confessional. Finally, there was the understanding that a physician had to be competent and accountable for medical interventions; it was presumed that there was objective expertise.

Between 460–377 BC, then, there was a code or theory of ethics that was based on the possibility of objective knowledge of the good of the human species, on the concept of what constituted well-functioning for the human organism or system, on the human function of reason, and on the interconnection of biology and ethics. This medical ethics tradition was transcultural and transtemporal because it was based on the global understanding of the human species and its place in the cosmos. It was not isolated in Greece because it was not foundationally an expression of Greek politics, but a worldwide profession of healing physicians. By 100 AD, Roman military medicine had incorporated this medical ethics and developed sophisticated surgery and infectious disease control. In medieval Europe, Dr. Michel Nostradamus, already licensed to practice medicine having successfully completed his degree at the University of Montpellier and answered a question on Hippocrates, returned from fighting the plague to complete his doctorate. He received a ring and the book of Hippocrates, as was the tradition. The Hippocratic code of ethics moved across the ocean to the Americas, modifying little with time or place. It continued in its basic form until what I call the Medical Ethics Discontinuity in the timeline, dating this somewhat arbitrarily with the publication of two books on a new medical ethics, one by Joseph Fletcher in 1966 (14) and the other by Paul Ramsey in 1970 (15). Rothmann places his "Critical Period of Change" between 1966 and 1976, so we are in fairly close agreement about the pivotal point in the change. What was overturned in the new bioethics at that discontinuity? I don't feel it is an exaggeration to say that world knowledge/world culture was overturned in medicine, as it was being overturned in other social institutions. I have selected the term "discontinuity" rather than change because of its paleontological meaning. At the major discontinuities in the history of life on this planet, most species became extinct, a few made it through the discontinuity unscathed, and some made it through modified greatly. These are major events.

Hippocratic ethics was not the only medical ethics that developed in the Roman Empire and continued to the discontinuity. When Christianity was Romanized, the world knowledge of the empire became incorporated into Roman Catholicism, with philosophic concepts of the importance of reason and free will, the natural order, natural law and the rights derived from it. Theology reinterpreted these concepts but did not reject them, and Thomas Aquinas made the bold leap of faith that reason was compatible with faith. From about 1000 AD, an extensive body of literature on medical ethics grew within this tradition, focused on the immorality of violating natural law by impeding a natural process. Since procreation was seen as a natural, divinely-ordered process, the ethical discussion of contraception, abortion, and multiple sex partners all trace back to this ethical view. Although there were absolute religious principles incorporated in this medical ethics, there was also an acknowledgement of the frailty of human reason in any absolute sense (but importantly, not in any probabilistic sense). The ethical reasoning of casuistry (trying to fit human realities into an absolute principle) and Double Effect allowed compassion as well as objectivity. (See St. Augustine who is more compassionate than given credit (16).) Very significantly for U.S. constitutional law and ethical theories based on rights, this ethical tradition emphasized moral or natural rights, a concept that survived both the Reformation and the Counter-Reformation but did not do well in the apocalyptic Radical Reformation. This objective base for rights has been eroded by the new bioethics, which transforms rights into legal rights, therefore into political power contingencies.

MOVING THROUGH THE DISCONTINUITY OF 1960–1970

Following Vatican II, Roman Catholic medical ethics experienced the same shift from objective ethics to political social activism, although recent papal teaching has identified this secular "modernism" as a threat to ethics. However, even this critique is couched in the philosophy of phenomenology/existentialism, one of the pillars of postmodernism. Still, natural rights and the implication of the innate worth of any human being does give some base for resisting the current politicized bioethics. It is a shaky base, however, because it is not secular and can be dismissed preemptively for that reason alone. Also, within this tradition, liberation theology has introduced political ideology on the local level and furthered the changes desired by the new bioethics.

The Hippocratic medical ethics has also been changed moving through the discontinuity. Rather than an ethics, Hippocratic ethics has been reduced to a series of medical codes of ethics which are culturally relative, an expression of the fashions, opinions and taboos of a particular time and place, and the influence of whatever political ideology is in power. These codes easily shift and reverse. For example, a political power demand for zero tolerance of any social relationship between a psychiatrist and patient, forever, has now replaced the more flexible codes of the 1950s (17). Code changes compatible with new economic care delivery systems have been made in the last two decades. These shifts are based on the relative political power of the participants in a negotiating process, and the possession of levers for group dynamics (coercion). Hippocratic ethics has now become irrelevant, made fun of, eliminated as an oath in medical schools, and replaced with professional codes of ethics that transform into standards of practice for the time, criteria for malpractice litigation. At that point, these codes are not ethics at all, but folded into a legalism.

I have partially discussed the factors leading to this discontinuity in chapter one. Although Rothman states the goal of the new bioethics quite clearly, he still feels there are prior causes for the political move, so I want to take a look at other candidates. Rothman suggests that the depersonalization of medicine produced the impetus for a major change, hence his title "Strangers at the Bedside." But he may have the cart before the horse. Until the new bioethics supported medical care delivery system changes, depersonalization does not seem a major issue. With the politicoeconomic move to prepaid insurance plans, changes did occur in medicine, and bureaucratic organization and regulation for cost grew. Still, economies of scale and group practices were not that prevalent. The bedside manner of physicians has not been demonstrated to have changed appreciably through the 1940s or even into the 1950s. After the new bioethics, however, in the 1960s and 1970s, physicians did change appreciably, and I am more inclined to think the former caused the latter.

Claims are also made that technology (in keeping with the political ideology that denigrates the "technological fix") produced the discontinuity. However, very major technological changes have occurred before in medicine: the microscope, the cell theory, vaccines, forceps, antiseptics, anaesthesias. These did not produce a major break with the ethical system of medicine, yet consider the major implications of all of them. These changes did not occur within the context of a major cultural break, however (social and intellectual factors such as: antiscience, narrative truth theory, ideological rejection of all expertise, civil and legal rights emphasis, expansion of the tort law system for advocacy, use of Nazi experimentation to fan anti-medicine, environmental apocalypticism, and the academic turn to postmodernism, all culminating in a cultural turn in the 70s). It is rather these conceptual frames that, intentionally, created the discontinuity. I could include other more economic factors, but factors closely related to or amplifying postmodernism: the already discussed third-party insuring bureaucracies and their move to prospective payment schemes, the limits-to-growth movement, vanishing internal markets in developed countries that require globalization, and the continuing centralization of government power (18–20).

The importance of academic theories should not be slighted in the discontinuity. In the next chapter, I will show how intellectuals were moving toward a relativism and narrative theory of truth (hermeneutics) and away from logic and empiricism – the move toward nihilism. This nihilism and political interpretation of reality characteristic of postmodernism replaced both deontological ethics and the naturalistic ethics of the line from Aristotle to Hume. Political might did make right.

THE BIOETHICS OF THE 1970s: AUTONOMY

The bioethics of the 1970s appealed to ideas of freedom to choose, rights, and patient moral autonomy, which quickly became political autonomy. Although the Kennedy Institute at Georgetown used Potter's term "bioethics" and may have appropriated the term without realizing its source, these bioethicists did not mean by the term a true interdiscipline of science and values. Instead, special moral reasoning and a political activist consensus were important to ethical analysis of medicine. Since such a postmodern ethics is time-and culture-bound, the new bioethics incorporated the social and political mores of the 1970s and its cultural revolution. Physicians were defined as facilitators of patient choice or as body mechanics doing what the customer wants (21) or as contract-makers. They were supposed to be value-neutral or were accused of "paternalism," the code word for an attack on the value system of medicine as embodied in the Hippocratic code of ethics. Issues were framed politically, often as the oppression by the white European male professional. The suggested remedy was a formalized method for obtaining informed consent which became, because of legal implications, defensive medicine; and an emphasis on civil or legal rights for patients. The Principle of Autonomy during the 1970s was the overriding value, and its bureaucratic implementation became consent forms and the legal use of state health laws describing medical treatment without patient consent as assault and battery, an indication for malpractice. Constitutionally, patients were given, through privacy rights, the rights to refuse treatment, have treatment withdrawn, or not be involuntarily hospitalized.

For a philosopher, autonomy sounds as if it should have a Kantian origin, with that philosopher's ethics stressing moral autonomy and the moral agent. However, Kant was politically reinterpreted so that autonomy meant a type of political liberalism rather than the strict rationalism at the base of all Kantian ethics (22). The moral agents of the 70s medical ethics were not the moral agents of Kant, who were required to meet standards of reasoning that many legally-competent patients would not meet. Autonomous choice seemed closer to Mill's political theory and ethical conclusion that the individual had the best knowledge of his or her good (forgetting that Mill too had a high standard for such competency to evaluate one's good), but the new bioethics also reinterpreted Mill to imply that no one else could hope to judge what was good or bad for a patient in a medical case and give an objective conclusion. It was assumed that any evaluations asserted to be based on standards of reason and empirical reality would be the imposition of another person's subjective, idiosyncratic values on the patient: paternalistic, judgmental, moralistic, coercive, idiosyncratic (23–26).

In the autonomy decade, the act of choosing was only restricted by the highly ambiguous prohibition not to take away the free choice of others. This is the only evil left possible under the new bioethics, since objective needs and interests such as existence (life), self-esteem, social bonding, and well-functioning are incapable of proof – but even this content for bioethics is arbitrary, as I have mentioned. It cannot be based in the postmodern ethical theory which asserts it. In reality, the act of choosing is heavily restricted within the new bioethics by the ideology and biases of the power group in control. For example, our university hospital had two cases of pregnant women, one raped while in a persistent vegetative state and one meeting legal criteria for brain death. The mothers of both women decided to keep their daughters' pregnancies continuing, and in the brain-dead case, this required life-support. Both mothers made their decisions after seeing ultrasound imaging of the fetuses. However, their decisions were criticized in print by a number of bioethicists who characterized the choice as treating the women as incubators, as denying their free (but unconscious) choice of an abortion. This is understandable when we considered that the Pro-Choice political position on abortion is the consensus position of bioethics. Not all choices are equal in the new bioethics, based on ideology. The same applies to a family or patient who chooses to continue medical care or heroic interventions, where ideology and cost-cutting economics within bioethics-approved health maintenance organizations severely restricts patient choice; and where bioethicists move from an autonomy principle to either a justice or medical futility principle arbitrarily.

Bioethicists also talk about patient autonomy or choice when discussing psychiatric involuntary commitment or medical refusal, as in Rivers vs. Katz (27). The standard for competency has been reduced significantly, with psychotic patients who were found to be requiring hospitalization under legal guidelines requiring much higher levels and proofs of risk to self or other and very severe illness, now also being found competent to make choices in certain areas in spite of this, such as taking medications. Examples of abuse of involuntary commitment and medications (and there are examples since psychiatry is not as scientific as it should be either in diagnosis or medicating) were used to make this change, and the argument was that competence was not a global or whole-person attribute, but was dependent on the question, competence to do what? A large class of individuals, already seen as discriminated against (and one lawyer has even used the term "sanism" to label that alleged discrimination) and worthy of protection, was moved from incompetence to competence. It was politically fashionable to be a defender of the mentally ill, so competence followed fashion. However, another class of psychiatric patients (outpatients), found themselves moved from competence to incompetence. Women going to psychotherapists for non-major therapy, who would be found legally competent under almost any standard of law and who would certainly be competent to refuse hospitalization or medication, were deemed by bioethicists not to have the capacity to consent to an affair with their therapist. A huge class of psychotherapy patients, generally but not necessarily women, found themselves moved to incapacity to choose, their autonomy removed for esoteric reasons such as "transference" which made no sense even under Freud's definition of "tranference-love," or for political reasons such as the assumed power imbalance or disparity between therapist and patient.

The Autonomy Principle of the 1970s, then, is a logically inconsistent principle with multiple contradictions. Generally, if a statement is contradictory, it becomes a meaningless or nonsense statement at face value. But there are hidden reasons why bioethics emphasizes the Autonomy Principle in some cases, and rejects it in others. I've given an example of the political ideology that is the real determiner of some choices for or against autonomy. Another determiner came to my attention personally.

I was a member of a hospital ethics committee when New York State passed legislation on Do Not Resuscitate issues. The statute gave patients the right to request that resuscitation not be done if there were cardiac arrest and respiratory failure. The New York State Department of Health has just issued its regulations interpreting the law (the most important part of health law is the "regs" that follow its passage), there were suggested forms for hospitals to implement the law and regulations, and our committee had met at the hospital to discuss the new DNR law. The meeting was lively. It was clear that under the DNR regulations, there was no requirement that a patient be terminal, that the family agree, that prognosis enter the picture, or that more than legal competence be required. The law did contain some provisions for a psychiatric evaluation, but not necessarily by a psychiatrist. Given time and staffing constraints, it was not likely that a very thorough evaluation could be done, and I considered this safeguard as meaningless given the actual operations of a hospital. The law basically gave a patient the right to refuse DNR and die, for no good reason that was required. Patients did not have to avoid risk and had the legal right to take life-threatening or life-destroying action in a hospital. The hospital counsel and I talked on the way to the parking garage, and in the intellectual concentration, I forgot to put on my seat belt, an unusual occurrence for me. Two blocks away, I ran into a safety check. I waited calmly, not realizing I did not have on my seat belt. The police officer came over, found everything with the car in order, but then noticed my seat belt was not on. He wrote me a ticket, because the State of New York had a compelling interest in keeping me from being hurt on its roads and required that I use the safety factor of a seat belt. I did not have a choice to take life-threatening or life-destroying action – on the highway.

The juxtaposition intrigued me. There was only one answer, which brings me to the next development in the history of bioethics. If terminally-ill patients, or any patient actually, is not resuscitated, the health care system saves many dollars in what would have been future expensive medical care. If a driver or passenger is injured in a car accident, many dollars will be spent for treatment, but if he or she is wearing a seat belt, statistically there will be less expensive injuries. New York State did not wish to have a compelling interest in my life that activated its commonwealth or police powers. It had a compelling interest in my cost. Autonomy would be the chosen bioethics principle primarily when it saved costs, and a modern version of utility would be the ethical principle when autonomy did not save costs but created them. The changes in bioethics of the 1980s had arrived.

THE SHIFT TO UTILITY IN THE 1980s–1990s

Originally, the new bioethics' universal ethical principles included the remnant of objective Hippocratic medical ethics in its universal principles of beneficence and nonmaleficence (do no harm). The universal principle of justice, however, with its emphasis on a "just" distribution of "scarce" resources combined with the politicoeconomic pressures to change the meaning of those principles. The current version of utilitarian theory focuses on social utility and cost-efficiency, something present from the very beginning of the new bioethics but not made explicit. Calabresi and Bobbit's work on economic first-order decisions and economic second-order decisions is an excellent description of the interface of ethics, politics, and economics (28). They show that most bioethics decisions are of the economic second-order type. A set amount of money is the given, the limitation that creates what they call "tragic choices." Within that constraint, ethicists, physicians, lawyers, and patients try to determine what is the best choice, given the limitation. For example, we had an AIDS case where the young man was admitted to emergency for pneumocystis carinii pneumonia. To survive, he would need to be admitted to intensive care where he could receive ventilator assist. At the same time, a man suffering a myocardial infarction (MI), his first, arrived in emergency and also needed to be admitted to the intensive care unit to survive, although his chances of survival were excellent and there was little damange to the heart. The man with AIDS had had previous PCP admissions and his final prognosis, before the AIDS cocktail drugs, was terminal. He had also signed a DNR request and prior directives about no extraordinary care, but now in shock that this might preclude him from the intensive care unit, told his physician he had not realized the full meaning of his prior directives and did not think it had meant now, here, death.

The ethical problem was that there was only one bed available in the intensive care unit, and at the time, no ventilators were available except in that unit. The medical resident in charge of the unit, the AIDS patient's physician, and the emergency room now had a very serious life-and-death ethical problem. No other hospital had a bed available in their intensive care units. Medivac to another region was not feasible either. It is hard to argue that the ethical decision should be made on a first come, first served basis, on a lottery, or on social desirability. But it is relatively easy to argue that the ethical decision will have to be made on likely outcome of treatment for both patients, and the likely outcome is far, far better for the MI than for the AIDS patient. Still, that is not the full ethical analysis, nor should it be. What is being discussed is a second-order decision only.

The more important ethical issue is on the first-order decision-making level, and an even more adequate ethical analysis would raise the question of why bioethics discussion appeared restricted to the second-order decision level, why a thorough look at first-order mechanisms was not being undertaken. Why was there only one available bed in the intensive care unit? What had structured the medical care system in such a way that it could offer only one bed in an emergency?

This analysis of first-order decisions takes us directly to policy experts and to politics as the base for ethical decisions in the real world. Most first-order decisions, Calabresi and Bobbit will argue, are not done by public vote, but by policy agencies and individuals whose names never reach the public. These shadow decision makers are appointed, not elected, and are assumed to make their recommendations/implementations in a professional, non-ideological, non-biased way. However, the very expertise they use implies value judgments made before the deliberations are ever made. One value judgment is availability of resources, and as work in limits to growth has shown (29), many assumptions about acceptable environmental quality, acceptable population growth, restriction of research, "appropriate" technology, and ethical risk choices or prudence are embedded in resource/growth projections – not to mention the frequent failure for such projections to hold in the real world. The most serious mistake made in such work is that it does not successfully factor in advances or new realities, but may make the mistake of projecting trends in research and availability into the future without computing for change in the model.

Resource allocation is also heavily value-laden, since it must prioritize and redistribute, and those activities are more often political than ethical. Much has been written about the excessive amount of GNP that medical care takes, but the term "excessive" is a value term. The ethical discussion has not taken place, however, because it is already assumed that whatever the percent of GNP, it is too large a percent. Resource distribution also makes undemocratic choices about what areas will receive the resources, and here the implied ethical theory finally becomes clear. It is a modern version of the utility and efficiency components of utilitarian theory, coupled with statistical morality. Quality medical care has been redefined as cost-effective or cost-efficient medical care, again without much bioethics discussion, because the politicoeconomic component of the new bioethics, accepted from the beginning, gained strength in the 1980s and now trumps the autonomy principle. It was, however, always present. It is for this reason that the autonomy principle was only honored when it was cost-efficient, that the nonmaleficence principle was used to restrict the use of medical technology as high-risk or harmful and furthered the downsizing of medicine, and that the beneficence principle became utility outcome studies. The needs of the individual patient, and the response of that individual patient, are conflated with the needs and responses of all aggregated patients (utility). This statistical morality appears to be quantifiable and scientific, but philosophers have long criticized it as arbitrary (30). The weighting is problematic. Comparison is difficult. The individual patient is sacrificed for the aggregate, but there is no rational justification for such a preferential selection or numerical weighting. From a systems view if one level of a system is consistently sacrificed, the entire system may collapse (31). Yet physicians, despite the known individual variability within a biological species, continue to embrace the utility/efficiency component of the new bioethics. They have become active in resource allocation and gate-keeping, accepting the imposed scarcity of resources as a natural rather than a political situation. They are very compatible with the statistical protocols and the new computer modelling of medicine.

There was a second aspect of the utilitarian bent of the 80s in bioethics. The principle of beneficence became understood as first the communal good, and then the commonwealth or police power of the state. Physicians have accepted the role of agents of the state for ideologically-appealing situations such as patient sexual abuse, child abuse, dangerousness, drug abuse, and battering. Extensions of the meaning of these categories, based on the idea of "health" have the potential to be endless, but bioethicists accept this extension because the health goal is another component of the ideology. Health may become a statist goal enforced by medical experts as part of bioethics (32). Because health risk is a social utility or efficiency concept, the means to deal with such risk is the orchestrated modification of the behavior of large populations. This "Health Belief Model" requires an individual to change behavior that may be of little risk to that individual on his or her level, but is unacceptable to society on the communal level. Because the term "health" is so undefined, mass population control may be done based on social fads, pseudoscientific errors, or political assumptions, and we are probably on T. Coraghessan Boyle's *The Road to Wellville*.

Hippocratic ethics, already collapsed to a legalism in codes of ethics, now was submerged as cost-efficiency and utility. Roman Catholic medical ethics, caught in the same second-order decisions for downsizing and preoccupied with the social gospel which required reallocation disputes between middle class and underclass, further lost its ethical base. The shamanic tradition still continued as an inexpensive alternative to modern medicine and could be used for cost-containment. It was also redefined so that it seemed to encourage less use of medical facilities and resources. "Natural" or "spiritual" treatments, because their standards are partly subjective or faith-based, need not be evaluated by scientifically-objective standards of good outcomes or meeting biological needs. Physicians are encouraged to make use of them or work with practitioners, with little emphasis on harmful side-effects or risk-assessment.

"Natural" is assumed to be "good." Bioethics, as the ideology of postmodernism and the cultural changes of the last three decades of the twentieth century, proved very compatible with this new form of utility/efficiency/futility medical ethics. Political and social ethical relativism was now firmly entrenched as the decision-making base and the content-less value base for the profession of medicine, whose professional standing was now eroded and threatened.

BUREAUCRATIZED, MANAGED POSTMODERN BIOETHICS

We have found that the timeline leading to a bureaucratized, manager-dominated postmodern bioethics was drawn from the assumption of a content-less ethics first called bureaucratic ethics by Engelhardt but found in practice in the proliferation of government task forces and committees on bioethics questions and the conflating of ethics with public policy and politics. This philosophic assumption rather than valid conclusion that no verified knowledge is possible in ethics implies that the only standard for ethical decision making can be the process itself, with the selected negotiators coming from multiple but all equally invalid ethical belief systems. Consensus and premature closure are the norm, and group dynamics reflect the power alignment of society. As psychiatry and psychology have shown, group dynamics can be manipulated by a skilled agenda-setter and chairman, and by their own internal forces. Seemingly free and democratic negotiations often aren't, yet this process became elevated over scientific verifiability or logic. Its resemblance to current political process is not coincidental.

In another recognition of this politicizing of bioethics, Brody suggested adding to the Kennedy Institute's litany of autonomy, beneficence, non-maleficence, justice and truth-telling the additional principle of Power (33). These cultural steps have converted bioethics to power struggles, the stuff of politics.

By the 1990s, bioethics found all narrative to be equal in the negotiating process, but the political ideology narrative was, of course, more equal than others. Bioethicists denied they could know any valid content within the process ethics, although preferred ideological content was imported. As postmodernism, everything was a "conversation" (34) and these conversations were occurring within Presidents' Commissions for the Study of Ethical Problems in Medicine and Biomedical and Behavioral Research, hospital ethics committees, professional subcommittees on ethical codes, federal and state task forces, and the call for ethical committee oversight of that last bastion of uncontrolled medical care, reproductive medicine.

Gradually, all the trends on the timeline converged. Autonomy and moral/natural rights theories of ethics had been converted into civil and legal rights, into law and medicine. Natural rights based in a natural world independent of social political organization were replaced by civil/legal rights based on political power and therefore capable of being given or taken away depending on the contingencies of power. The abortion debate is a prime example of the confrontation of natural/moral/human rights (pro-life) with civil/legal rights (pro-choice). So far, this legal and political sense of rights has been dominant, with rights being political constructs relative to time and place, or nihilistic since they are brokered by power groups in a specific society. The euthanasia debate is also an example. Rights to refuse expensive medical treatment are easily guaranteed by law, while rights to expensive medical treatment are nonexistent except in rare cases such as emergency treatment or treatment for involuntarily-committed patients. Some may protest that bioethics wishes to give every individual the right to access to medical care, but the delivery systems that will distribute that guaranteed care ration the care available. The actual right becomes the right of equal access to unavailable care for everyone. This right also runs into the medical futility question, where, unlike its other attempts to control medical decision making, bioethics here argues that only the professional can decide if care is appropriate, so that a decision of medical futility counters any patient's request or right to treatment.

The legal and political lines merged with the economic cost-containment goal and the population manipulation role of physicians, once any ethical or legal right to treatment for an individual had been negated. There is now no legal or ethical base to safeguard a patient's need to receive care, and the managerial class determines what care can be accessed. Bioethics does not restrict the managerial class in its decisions about care, except in those cases where the delivery system is for-profit, since judging capitalism to be unethical is part of the bioethics political consensus. There is no difference between the management decisions of non-profits or for-profits, but bioethics has only criticized the for-profits to any extent. Nor are governmental managerial decisions seen as potentially unethical, since there is often a preference for government third-party payers over multiple third-party payers.

Going into the twenty-first century, what we have is a bureaucratized, managerial-based, cost-efficiency aimed, ideological system for bioethics and medicine, a postmodern bioethics and a new name for medicine: the health care system. This change of name is significant and I will discuss it in more detail in Part II. This new bioethics and change in the medical system is comfortable to young physicians. First, it reflects the opinions of their decade. It is culturally relative and they are most comfortable with either a culturally relative ethics or a content-less ethics, since they do not wish to impose their beliefs on anyone else and are reluctant to make evaluations. It is a wishful belief in democratic process and conversation and fits the political ideology of many physicians. How democracy is defined is not their concern, and the definition easily slides into a feelgood, "can't we all just get along" pluralism that eschews outliers, confrontation, or passion. They are comfortable with compromise and consensus. It is expressed in seemingly clear legal statutes that order a specific action and remove, they believe, the ambiguity of ethics. Most physicians first ask what options are legal, rather than what options are ethical. There is little discussion of philosophy of law, or good laws and bad laws. That implies opposition and a darker, deeper view of politics, of realpolitik or power politics. It allows physicians to cooperate with whatever economic allocation choices are made by third-party payers and policy planners. This gives a justification for what they see around them in the health care system, cost-efficiency and downsizing. Second-order tragic ethical choices become more palatable for them. It has consensus-building style that resembles the psychotherapeutic model so prevalent in American society. This model also makes all decisions the patient's responsibility and accountability, not the physician's. Given the enormous emotional cost of physician accountability, it is a relief to follow the therapeutic model and burden the patient with ethical accountability.

I was asked in 1996 to project a scenario for bioethics into the next 25 years, and the title of my article was that it was not a pretty picture (35). Based on the historic timeline I have presented here, and expected continuing or changing trends, as well as "reports from the field," I came to the following conclusions.

Twenty-five years from now, medicine is likely to be more and more bureaucratic and impersonal. It too is being converted into a civil service model. Cost-efficiency systems will eliminate continuity-of-care. The physician will be just another health provider or service provider.

There will be growing anger on the patient's part, and growing frustration on the physician's part. The two are being forced into adversarial positions over fundamental questions of pain, loss, living, and dying. Trust and respect will erode and be difficult to find.

Throughout the developed world medical systems may approach the inadequate levels of the current Russian system collapse. There will be a two-tiered system, with the politically powerful surreptitiously getting the benefits of past research and application, but since there will be little new research and development, even the powerful will find themselves expendable.

The middle class and working poor will be on their own. Ethicists will claim we can no longer afford to care for the retired, demented, mentally ill, chronically ill, and terminally ill at the levels we once did – or not at all. This will be presented to the public in more palatable ways through continued indoctrination. In the name of bioethics, there will be age cut-offs increasing, withdrawal of care without request the standard, and no care for premature babies below ideal protocol weights.

Public health will initially be cut back and traditional measures will not be used because of political ideology. Vaccines will be used only if demonstrated to be cost-effective. This will conflict with a strong trend toward severe outbreaks or epidemics as microbes naturally become emerging, resurging, and resistant. The response, however, may be public relations management or blaming the human species for environmental degradation. We could hope that a potentially good consequence from future epidemics would be forcing public policy makers to see the practical good of an adequate medical care system, modifying the downsizing of the medical institution. That may be optimistic.

We will have to be vigilant 25 years from now to prevent physicians adopting the role of agents of politicoeconomic-pressured terminations. The ethical battle will be over saving the role of physicians, and holding the line against economically-driven "medical futility" and physician-assisted termination. My worst project is that economic forces will be so powerful, objective and rational ethics so neutralized, and patient termination so spin-doctored with words like "choice," "dignity," or "duty," that the health care system will terminate patients based on social utility criteria no longer disguised as patient autonomy or medical futility. The power struggle over ethics and politics has occurred before in medicine, and physicians may lack the ethical means to come to the ethical end.

A basic ethical question will be the science of human nature. What is the normal functioning of the human biological system? We can use the new information from the Human Genome Project to contrast normal good functioning from extreme variance. However, if medicine is too politicized in 25 years, this ethical exploration will be censored and replaced by the ideology of power groups. The promise of microbiology and biotechnology may have to struggle against the ancient and powerful forces of thought control in postmodernism, and could easily lose. The ethical issue of the biological significance of parenting and family may be decided by the trend toward depersonalization, which denies biological bonds and weakens family and social bonds. Physicians trained in terms of autonomy, individuation-separation, and dysfunctional families could become agents for this depersonalization of social life in the name of expertise. The supportive, personal social net will be threatened. The family will continue to be shunted aside in the health care system, both by health law and bioethics. That exclusion will leave the sick patient to stand alone against the tenets of utility, futility, and quality of life.

Physicians, more and more salaried and in managed-care systems, will have great difficulty resisting efficiency, utility and total quality management standards. Bioethics will have re-educated them to accept those standards, and if they can transform themselves into civil servants, they will. If they struggle to remain classic physicians, there will be frequent crises of conscience and economic repercussions.

The ethics of abortion will be revisited as the cognitive dissonance of imaging the growing fetus earlier and earlier in photographic detail, and culturally calling the fetus a baby except in the abortion context, clashes with the demand for the civil absolute right to choose to terminate the pregnancy and kill the fetus. This political confrontation between an absolute right and respect for life cannot be resolved only by Supreme Court decisions that lack a sufficient consensus in society. New abortifacients that cause very early abortions may alleviate some of the problem, particularly since very early abortions have more of a social consensus than mid-term and almost full-term abortions. Physician involvement would be lessened, requiring an office call, prescription, and monitoring, and therefore eliminating the growth of abortion clinics. Couples may want to make decisions to continue or terminate pregnancies where a genetic variation exists, and there will be ethical arguments over what genetic variances can justify termination. Physicians, in a wish to avoid this issue that is often placed in ideological frames, may arbitrarily and unfairly restrict genetic testing.

Restricting care, or following by-the-book protocols, may be the common standard of practice. The clinician will have less and less power to make decisions, with statistical medicine and morality dominating. Bioethics, reflecting economic and political trends, will support this.

And lawyers will always be with us. Tort law is not likely to be reformed.

For physicians and bioethicists, what Norman Maclean wrote in *Young Men and Fire* is a vital warning: "I see better what happens in grass than on the horizon. Most of us do, and probably it is just as well, but what's found buried in the grass doesn't tell us how to get out of the way." (36).

REFERENCES

1. Kelly, D. R. (1979). *The Emergence of Roman Catholic Medical Ethics in North America.* New York: Edwin Mellen Press.

2. Rothman, D. J. (1990). *Strangers at the Bedside: A History of How Law and Ethics Transformed Medical Decision Making*. New York: Basic Books.
3. Macklin, R. (1993). *Enemies of Patients: How Doctors Are Losing Their Power and Patients Are Losing Their Rights*. New York: Oxford University Press.
4. Engelhardt, Jr., H. T., & Wildes, K. W. (1999). In the beginning: The emergence of secular bioethics. In: R. B. Edwards & E. E. Bittar (Eds), *Advances in Bioethics* (pp. 1–16). Stamford, CT: JAI Press.
5. Clements, C. D. (1992). Systems ethics and the history of medical ethics. *Psychiatric Quarterly, 63*(4), 367–390.
6. Clements, C. D.(1994). Systems bioethics: AIDS, reproduction, and the pandemic destroying population groups. In: E. Bittar & N. Bittar (Eds), *Principles of Medical Biology*, Vol. 1A (pp. 88–114). Stamford, CT: JAI Press.
7. Clements, C. D. (1999). Systems bioethics: The future of epidemics and medical ethics. In: R. B. Edwards & E. E. Bittar (Eds), *Advances in Bioethics* (pp. 31–60). Stamford, CT: JAI Press.
8. Dioszigi, V. (1968). *Tracing Shamans in Siberia: The Story of an Ethnographical Research Expedition* (Translated by Balbo, A.R.). New York: Humanities Press.
9. Burden, G. (3/7/2000). 'Doc' like an Egyptian: Ancient Egypt was a think-tank for medicine, as a tome from the 13th century uncovers. *The Medical Post, 36*(9). Toronto: RM Publishing.
10. Levi-Strauss, C. (1968). *The Origin of Table Manners: Introduction to a Science of Mythology*. New York: Harper and Row.
11. Winfield, M. (1997). *Shadows of the Western Door*. Buffalo, NY: Western New York Wares.
12. Clements, C. D. (1983). The lesion and the function: Setting up the reductionist problem. *Perspectives in Biology and Medicine, 26*(3), 433–440.
13. Pellegrino, E., & Thomasma, D. (1981). *A Philosophic Basis of Medical Practice*. New York: Oxford University Press.
14. Fletcher, J. (1966). *Situation Ethics: The New Morality*. Philadelphia: Westminster.
15. Ramsey, P. (1970). *The Patient as Person*. New Haven, CT: Yale University Press.
16. Wills, G. (1999). *Saint Augustine*. New York: Lipper/Viking.
17. Gabbard, G. (Ed.) (1989). *Exploitation in Professional Relationships*. Washington, D.C.: American Psychiatric Press.
18. Meadows, D. H., Meadows, D. L., Randers, J., & Behrens, W. W. (1972). *The Limits to Growth*. New York: American Library.
19. Jecker, N. S. (1999). Justice in health care. In: R. B. Edwards & E. E. Bittar (Eds), *Advances in Bioethics* (pp. 285–299). Stamford, CT: JAI Press.
20. Morreim, E. H. (1991). *Balancing Act: The New Medical Ethics of Medicine's New Economics*. Dordrecht, The Netherlands: Kluwer Academic Publishers.
21. Bayles, M. (1978). Physicians as body mechanics. In: J. W. Davis, C. B. Hoffmaster & S. Shorten (Eds), *Contemporary Issues in Biomedical Ethics* (pp. 167–178). Clifton, NJ: Humana Press.
22. Kant, I. (1959). *The Foundations of the Metaphysics of Morals* (Translated by Beck, L. W.). Indianapolis: Bobbs-Merrill.
23. Veatch, R. M. (1981). *A Theory of Medical Ethics*. New York: Basic Books.
24. Hayry, H., & Hayry, M. (1990). Liberty, equality, utility: Classical to liberal utilitarianism. In: T. Campbell (Ed.). *Law and Enlightenment in Britain*. Aberdeen: Aberdeen University Press.
25. Culver, C., & Gert, B. (1982). *Philosophy in Medicine*. New York: Oxford University Press.
26. Beauchamp, T., & Walters, L. (Eds) (1978). *Contemporary Issues in Bioethics*. New York: Wadsworth.

27. Ciccone, J. R., Tokoli, J. F., Clements, C. D., & Gift, T. E. (1990). Right to refuse treatment: Impact of Rivers v. Katz. *Bulletin American Academy of Psychiatry and Law, 18*(2), 203–215.
28. Calabresi, G., & Bobbit, P. (1978). *Tragic Choices: The Conflicts Society Confronts in the Allocation of Tragically Scarce Resources*. New York: Norton.
29. Meadows, D. H., Meadows, D. L., Randers, J., & Behrens, W. W. (1972). *The Limits to Growth*. New York: American Library.
30. Brock, D. W. (1973). Recent work in utilitarianism. *American Philosophic Quarterly, 10*(4).
31. Pattee, H. H. (1973). *Hierarchy Theory: The Challenge of Complex Systems*. New York: George Braziller.
32. Jeffery, R. W. (1989). Risk behaviors and health:Contrasting individual and population perspectives. *American Psychol, 44*, 1194–1202.
33. Brody, H. (1992). *The Healer's Power*. New Haven, CT: Yale University Press.
34. Oakeshott, M. (1991). *Rationalism in Politics and Other Essays*. Indianapolis: Liberty Press.
35. Clements, C. D. (1996). Medical ethics in 25 years – not a pretty picture. *The Medical Post, 32*. Toronto: RM Publishing.
36. Maclean, N. (1992). *Young Men and Fire*. Chicago: University of Chicago Press.

3. THE POSTMODERN ENVIRONMENT OF BIOETHICS

THE CONCEPTUAL TRIPOD OF POSTMODERNISM: PSYCHOANALYSIS, EXISTENTIALISM/ PHENOMENOLOGY, AND NEOMARXISM

A few years ago, papers carried the news that Jean-Francois Lyotard, the French philosopher, had died at age 73. Lyotard was one of France's leading postmodern philosophers, and his obituary recalled that his philosophic work was influenced by the 1968 student-worker riots in Paris. That was an important statement, because it showed the strong political assumption of postmodernism, its neomarxist ideology, expressed both in the U.S. and in Europe and centered on the student/academia activism of the 1960s and 1970s.

The obituary went on to list Lyotard as one of the most respected figures in French philosophy, along with Gilles Deleuze, Michel Foucault, and Jacques Derrida, in the second half of the twentieth century. All were postmodernists. Lyotard's work, *The Post-Modern Condition, The Differend, Heidegger and the Jews, Lessons on the Analytic of the Sublime,* and *Discours, Figure*, described as embodied in the 1968 riots, focused on the individual against the establishment. But that would be a misleading phrase unless the political ideology behind the student movements of the 1960s and 1970s were understood. The New Left, or neoleft, or neomarxism supplied the ideology and control of those movements, so "the individual" must be understood as the politically-correct unit of a cohesive group whose goals and strategies overrode the individual human being. If we interpret postmodernism as based on the natural rights of man or on Enlightenment liberalism, we will go astray. It is nothing of the sort.

The French paper Le Monde, in a few short lines, actually sketched the conceptual framework of postmodernism quite well: "He belonged to that beautiful generation of French philosophers for whom the works of Freud,

Marx and Neitzsche would play an essential role after the events of 1968." Here in the briefest form, is the Conceptual Tripod of Postmodernism. Psychoanalysis is represented by Freud, and it is one of the important bases of postmodernism. Marx was modified slightly into neomarxism, but Marxism has always been a fundamental for postmodernism. Nietzsche, uncomfortably reinterpreted, has to represent Existentialism and Phenomenology, and this is not totally unfair because Nietzsche was one of the bases on which Existentialism and Phenomenology drew in going in a direction that Nietzsche himself would not have approved. These three legs of the postmodern tripod all share common ideas.

Freud, though a scientific reductionist as I think Sulloway has demonstrated (1), put aside criteria of logic, reason, and neurology in order to try to understand the patient's ideas and emotional responses. There were multiple realities in his patients' thought-worlds, and psychoanalysis was not concerned with distinguishing between reality and fantasy. For therapists, what the patient says (conversation) is more important than reality, and for some therapists, reality is only one of many stories. Aryeh Botwinick, in his book *Postmodernism and Democratic Theory* (2), sets up the core tension in psychoanalytic theory by quoting Yankelovich and Barrett. They agree with Sulloway that Freud's abstract theory, its metapsychology, is a pure form of scientific materialism, but that psychoanalytic practice is existential at its center, so that psychoanalysis attempts to be both in scientific materialism and in existentialism, a serious contradiction within the theory. Botwinick suggests that the concept of the unconscious, seen as story or theme, redefines rationality as "power." Human nature is not seen as scientifically biological or ethological, but as metaphoric. Postmodernism interprets psychoanalysis primarily as the therapeutic function rather than the theory, with individual therapeutic acts constituting the real theory of psychoanalysis. This allows Botwinick to say that: "Psychoanalysis . . . is 'postmodernism' in advance of the historical advent of postmodernism." I think he is quite right to see the therapeutic model as dismissing objective reality, reducing the human-nature interaction to story or conversation, and allowing fantasies about ideology to replace verification or correspondence concepts of truth or pragmatic reality. What for Freud was a practical dualism between neuroscience and psychology until the scientific reduction could be done becomes talk, metaphor, assumption, and finally fantastic ideology.

Existentialism and phenomenology are closely-linked philosophic positions that form the second leg of the postmodern tripod. Existentialism denies the objective reality of a human nature as an essence that precedes existence and maintains the personal creation of what one is. It is also hermeneutic in its development, more wed to a coherence concept of truth rather than an empirical science correspondence to a reality independent of man's stories and authentic choices. Phenomenology is a partial extension of Kant's argument for a transcendent ego and conceptual constructs (time, space, causality, substance) that order or structure natural experiences. With Heidegger, the products of different time periods and histories are not commensurable in any full sense, so concepts and narratives are culture-bound and not transferrable. The possibility of an ongoing world culture or world knowledge is then illusory in any sense of understood content. Botwinick talks about the postmodern interpretation of democracy as the institutionalization of a generalized agnosticism (what I will call an empty process). Theorizing is separated from its objects, so there is no protection against our uncertainties (or freedom, in the existentialist sense), and we have left only our continually reactivated moral intuitions and alarm systems (our existential choices that create ourselves), which Botwinick assumes is all we ever have anyway. What this means for ethics and all value systems is, as he asserts, that the moral, social, ethical, political, and aesthetic ideal of "tradition" changes to "sincerity" and then "authenticity," the fuller self-consciousness of the methods and goals of the intellectual life. The term "tradition" covers the previous base of reason and scientific reality. Sincerity or authenticity replaces the external sources of authority (read, reason) and value (read a naturalistic or objective ethics) with the individual who is the sole guarator of the genuineness and seriousness of his or her commitment, argues Botwinick in a straightforward existentialist position.

Existentialism and phenomenology have never been compatible with the self-correcting scientific method of ascertaining probable truth, but have made use of some scientific theorizing such as quantum physics which appears to those who hold these positions as compatible with the coherence standard, narrative concept, and personal mentation base of these philosophic schools (3). The activism of a 1995 conference at the New York Academy of Sciences, "The Flight from Science and Reason," may be unfortunate, but when participants identified the exploitation of scientific ideas to attack science, they were on target. The physics of relativity and quantum mechanics, they argued, are sometimes distorted into conclusions that nothing in science is certain (and that certainty is a prerequisite for verified scientific knowledge, so there is no scientific knowledge), and that other stories such as mysticism, magic, and ideology have equal claim to belief.

To some extent, the sciences in their quest for completeness have brought this on themselves. Although not built into the scientific method, there is a tendency to think of perceived segments of reality as the limits of reality. A good systems methodology easily corrects for this, since levels of perception and levels of organization imply a perhaps endless reality, some of which science can grow to access, but all of which is highly likely to remain a moving goal. Kant, who was a good realist, did not think it was possible or required to know the thing-in-itself for science to accumulate knowledge. Nor is the goal of the scientific method to achieve omniscience. What science does is correct for human bias, human wishes, human fantasies in the understanding of some of the reality that has always been external to our biases, wishes and fantasies.

That such a corrective method is importantly required is shown by a small victory by Alan Sokal, a physicist. Sokal wrote a nonsense article that appeared to argue that the physical world did not exist. He criticized scientists who clung to the dogma of the Enlightenment that held sway over Western intellectual outlook until recently, the dogma that maintained an external world existed whose properties were independent of any individual human being and humanity as an entirety. His argument was purposely a gibberish he put together that contained no logical or scientific foundation. The article was accepted and published by a neoleft journal of cultural studies, *Social Text*, whose editors did not realize that the article was a parody. It was part of an issue devoted to the split between scientists and cultural critics of science. Following his small experiment, Sokal then wrote about his hoax for another journal. Before dismissing this small effort as no more than a joke, we should consider what actually occurred.

The article did not have to meet any standards of reason or empirical science to be accepted, and in fact, no such standards are viable within the philosophies or politics of postmodernism. The article's conclusions satisfied the prejudices of the mission of the journal. The attack on the existence of a world external to our minds is a fashionable one, and even if illogical or nonsensical, will be taken as a story of equal standing with any other story. Power and politics will determine expertise and our constructs of reality. All the products of human thinking are equivalent to a meaningless gibberish sent to a journal in order to prove a point. We could complain that Sokal was inauthentic and insincere, but according to the postmodern theory, only Sokal is privy to the knowledge of his authenticity or sincerity. We will have to accept on faith what he wishes to tell us about that.

Marxism or neomarxism is the third leg of the postmodern tripod. Karl Marx defined ethics as the imposition of the capitalist's perspective on all other classes, since he felt the economic establishment determined ethical writing and views. He really is a reincarnation of Thrasymachus and the argument with Socrates. Ethics was economics, not objective knowledge of the good of man. In his goal of a socialist classless society, ethics became human dignity, the healing of alienation. Ethics was also the political tool to achieve the classless society, politicized by definition. Neomarxism reinforced that ethics should be used as a political tool, should reflect the proper political consensus. In addition, ethical theory was a narrative that must include all oppressed voices, with those voices defined as oppressed having more power that those defined as the oppressors. Again, there is no objective truth on which to base this ethics, only conversations with groups given separate political evaluations or worth. Nor could logic or reason be of use, since traditional logical constraints are removed.

Botwinick makes the neoleft ideology quite clear (4). He finds the dominant political events of the 1990s to corroborate postmodernist thought, the convergence between contemporary liberal and communitarian thinking (as opposed to older constitutional liberalism) and the emergence of "democracy" as a monolithic political tradition of modernity. He talks about the asymmetries between capitalism and democracy and the trend to design a democratic political economy focused on decision-making power for the worker. Even the global economy and global political organization is foundational for this neoleft ideology. Regional economic associations and markets, with participatory activity on local matters but bureaucratic power on global matters, reflecting the Green movement slogan, "Think globally, act locally," should describe politics.

Even abstract ideas are seen as historically-rooted and dialectically suppressed in the class and power struggles that define culture. America is seen as part of a common legacy of issues uniting socialist and liberal-democratic societies worldwide. Old constitutional doctrines such as checks and balances which limit the amount of government-centralized power need to be discarded. Even old values such as individuality, diversity, and freedom of expression cannot be any longer supported, but must be replaced by a participatory (read activist) praxis that may or may not be an illusion of freedom (I would argue it is an illustion, the mob democracy that the classic Greek thinkers understood guaranteed no one freedom).

In keeping with this total politicizing, this total society, postmodern neoleft ideology seeks the continual diminishing of the social and the expansion of the political strategies and argumentation. Botwinick describes Laclau as saying (5). On a personal level, the self cannot stand aside, but must actively participate with others in large scale interactions in order to extend one's sense of self and not become impoverished and stunted. The only worthwhile individual must be an activist for the political ideology or consensus. Passive acceptance of this total society is not acceptable any more than active resistance is. Botwinick asserts that the ideology of this new left in the 60s, a radical deviation from the existing institutions, created a new tradition, one that left and should have left no important area of life outside its ideology, a very succinct definition of the total society, of totalitarian ideology.

The new left ideology also assumes limits to growth, assuming that economic growth cannot continue. Connor lists the concerns of a cultural politics of this sort: issues about race, gender, nuclear and green politics, the aged, and animal rights (6). He talks about the female as the patriarchy's Other and feminism which represents postmodernism. Cultural politics also talks about colonialism and post-colonialism in neoleft terms that deride the unified destiny of Man (the Western myth), a ruthless expunging of local or national histories and assumptions in the drive to universal rationalization and industrial progress, of universal history. This ideology has victimology or "victimage" as Connor calls it as a central ethical imperative because of its usefulness in taking the moral highground. Connor does, however, see the problem of the baselessness of these postmodern assumptions or values, of the inability of the position to justify democratic negotiation rather than force.

To the professional who is not a philosopher, or even to philosophers in bieothics who do not wish to see the implications for their field, this conceptual tripod, the discussion of the conceptual tripod, must seem unwarranted. The conceptual tripod of postmodernism is arcane, much more arcane than the brass tripod that Dr. Michel Nostradamus used in his locked study at night, studying astrology, alchemy, and the levels of being of the classic world. Dr. Nostradamus was both a shaman and a scientist, and very much rooted in an external world that was made up of complex levels that reason and experimental questionning could partly know. That Dr. Nostradamus referred to the ancient texts and that hermeneutic tradition, that he was a forerunner of the Radical Reformation, did not mean he saw the world as political ideology, as it does for postmodernists. In spite of its arcane quality, however, it is important to understand postmodernism in a general sense, if we wish to understand the new bioethics.

We could look at the implications of postmodernism from a clinical perspective, and there is no shortage of cases to illustrate the irrationalism of postmodern bioethics. There is the Baby L case in which a mother asked for continued treatment of a damaged infant, very much like the Florida Teresa case where the parents asked for continued life-sustaining care for their 12-year-old daughter who may have been brain-dead. The physicians in both cases wanted to discontinue treatment of the basis of medical futility (7, 8). In these cases, bioethicists argued the choice should not be respected, in direct contradiction of their arguments that physicians should always respect a competent patient or surrogate's rights to refuse treatment (9). I am familiar with an AIDS case in which the HIV-positive physician had requested full resuscitation and all possible treatment and not a DNR status, but the ethics committee was willing to consider going against his clear wishes/prior directives and placing a silent DNR in his chart. Since he was part of the hospital HMO system that would not cover him if he transferred to another hospital, the economic issues were seen to be relevant to the decision to quietly go against his will (10). But postmodern "conversations" do not have to be consistent or non-contradictory.

This postmodern conversation-model of reality, with its base a narrative standard rather than a validity standard, is fundamentally nihilisitc about the possibility of rational and objective bioethics and extremely agnostic about the possibility of knowledge in any scientific or logical sense, as we have seen (11–14). This basic irrationality is the reason that current bioethics can be arbitrary and inconsistent. For example, not only can patients have the right to refuse medical care, but they can also be denied the right to request medical care. Not only can patients have the autonomy to ask for physician-assisted suicide, discharge from a psychiatric hospital to the street, or abortion up to full term, and be granted it, but at the same time patients will not have the autonomy to get certain tests and treatments, experimental medical care, or even resuscitation. Bioethicists can choose the principle of economic utility as overriding the needs and choices of patients because they agree on a political principle of "just" distribution of resources that suits the ideological consensus (15).

Physicians are instructed by bioethicists to inform patients of every aspect of their illnesses, cold and harsh descriptions of treatment options to abrupt and blunt prognosis of death, because the principle of truth-telling is operative at end-of-life decisions and no one is particularly concerned about the emotional impact or the coercive nature of the information. Those same bioethicists, however, argue to delay implementation of genetic testing for diseases such as Huntington's, Cystic fibrosis or other late-onset genetic illnesses, and are strong advocates for laws and regulations requiring extensive pre-and post-counseling for HIV testing based on the principle not used for other decisions, that such information will be stressful (16).

Individual choice/autonomy/privacy is easily put aside for campaigns to manipulate large populations where alcohol consumption, tobacco smoking, weight control, hyperactivity or other "bad" or "unhealthy" behaviors are concerned, while the very same postmodern bioethics opposes campaigns to inhibit free sexual behavior (for example, abstinence and social disapproval of multiple sex partners), even when the risk of STDs and AIDS are apparent. There is no internal consistency or coherence within postmodern bioethics except for the arbitrary standards of power and ideology (17, 18).

In the next section, I want to show why the current bioethics denies the possibility of valid knowledge, ethics, beauty, and the importance of an individual unless that individual is a unit or number in an acceptable power group, or has some economically-useful function. In summary, postmodern bioethics denies that a group is governed by any standard other than ideology, with its utility and efficiency protocols. It denies that there is an objective reality external to the culturally-produced human mind, and asserts that all we can experience are cultural stories that compete with each other for power and control. It attempts a conversion of biology (medical science and bioscience) into politics and power strategies. All becomes political. Even in my work on AIDS/HIV, bioethicists wished to see this autoimmune deficiency disease as somehow a political disease to be cured politically.

IMPLICATIONS FOR REALITY, PROBABLE TRUTH, HUMAN WORTH, HUMAN FREEDOM

What is all the fuss about, a practical clinician might still ask. The historian, David Hollinger, in reviewing Gerald Holton's book, *Science and Anti-Science*, actually asks the same question (19). He describes Holton's "polemic" and his call to battle against the "irrationalist monster" unleashed by opponents of a science-centered world view. In fact, he himself submerges science into the opponents' picture of epistemology by talking about Holton being a champion of a particular notion of science's *cultural significance* within the *drama* of modern *history*. Hollinger's description of the targets of Holton's attack is also instructive, because he pictures them as "thinkers of our own time who doubt the continued adequacy of the classical Enlightenment construction of the relations of reason to freedom and human well-being." Actually, both Hollinger and Holton do share the same political ideology, which makes the review ironic. However, the key concepts are disturbing to the scientific epistemology. Hollinger reduces science to just another story in the history of culture, and admits that postmodern thinkers have altered the concept of reason in order to center on "freedom" and "human well-being." That these latter concepts have a contemporary consensus meaning in political ideology is not admitted, but is nevertheless accurate. The fuss about postmodernism is important, because it is a disagreement about how the mind works rationally, and about how the human organism interacts or transacts with the external realm independent of the human mind or culture – or one could add, political ideology.

Here are some of the key ideas that make postmodernism so fantasizing and pragmatically risky:

1. As we have seen, postmodernism collapses the natural external world into the historical and internal world. Oakeshott argues that there is no dichotomy between history (what men do and make) and nature (an external order). Nature becomes a fluid category subject to human design and control, anything we wish it to be (20, 21). Human stories about experience are internal to the human species with no posits or inferences possible about change, surprising new experiences, stubborn opposition to human desires. Human stories (and humans) cannot be explained except in a circle, since they are removed from any world of experience apart from themselves and must be self-posited, a new version of the old sui generis. The science of human beings becomes impossible to do, except at the level of literature, myth or ideology, where there are no standards. We can arbitrarily label political dissidents as having a mental illness, or use pseudopsychiatric diagnoses to campaign politically against opponents. We can eliminate all distinctions between normal and abnormal that we please. We can insist on Zero Risk or Zero Tolerance, without worrying about the profound risk/benefit consequences of those demands. We can convert environmentalism into a quasi-religious story with its own apocalyptic tradition. In a profound way, this makes the biosciences and medical science impossible, except as random, culturally-conditioned speech and writing.

2. Postmodernism disconnects human symbols from the natural world, elevating the symbols to the only "reality" or to "multiple equal realities" (22, 23). Botwinick defines reality as one possible ordering principle among many. Derrida uses the concept "differance" to allow meaning and objectivity to be displaced and deferred, rendering the idea irrelevant. As a result, there are only subjective perceptions, which are actually wishes, fantasies, stories, or ideologies. Corrective testing for bias or mistake about perception and experience becomes impossible, without a base; and the scientific method, as a result, is eliminated. In bioscience and medical science, then, the testing and research done cannot really claim superiority over any other perceptions or experiences, since all are basically untestable and arbitrary.

3. Postmodernism rejects generalizations based on probability, the use of categories tested in the world of experience, any encompassing theory, and any verified ordering of experiential impressions (24, 25). Experience becomes an unordered "buzzing confusion" with no possibility of using induction to understand it, and no explanation of the buzzing better than any other. Scientific induction is made inoperative, just another bias or artful design which humans are mysteriously prone to produce. This is the basic flaw of Kant's epistemology and metaphysics. He avoids the buzzing confusion of experience by postulating the objective (for him) categories in which humans process experience. Though he wishes to believe these categories are objective,

because his ego and his category are both transcendent, there is no way to validate or understand either that qualifies them as superior to other symbols or human designs. Not only are we incapable of knowing the thing-in-itself (which makes sense), but we cannot know the ego or category-in-itself or in any other way, which allows for no sense.

4. Postmodernism rejects the standards of noncontradiction, identity, consistency, testing in the world of experience, all of classic logic and reason (26, 27). Oakeshott claims there is no truth to be discovered and that reasoning is neither sovereign or alone. He advocates participatory conversation in which everything is permitted which can get accepted into the flow of speculation, everything is taken at face value, and acceptance replaces truth. This clear statement of the irrationality of postmodernism is somewhat modified by appeals to modal logic or a new multivalent logic, but more commonly, reason is simply discarded as unimportant to the achievement of human needs and interests (which are also highly subjective and incapable of rational or objective determination or explanation). What postmodernism has discarded are the basic standards for rational and perhaps even sane thought and interacting with the environment. It really is impossible, under this reading, to differentiate between average normal ideation and psychosis, between pragmatic validation of hypotheses in the real world and fantasies or hallucinations, between perceptions consistent with feedback from experience and perceptions that are off the mark. All cognitive standards are replaced with either individual preferences or with power. There would be no problem with therapists suggesting false memories of sexual abuse, and families, as a result, being destroyed. Therapists could testify as experts in the court system even though their interviews with children would be suggestions to the children that they had been sexually abused. Research logs could be falsified in order to get the results of the predetermined or desired conclusion.

5. Postmodernism distorts pragmatism to mean feeling good, having wishes, the individual believing that something is working for him and the view he holds, or, most important, to participating, acting to achieve wishes or fantasies (praxis) as Rorty would describe it (28). Human stories, cultural stories, must become relative to the idiosyncracies of the individual who wishes to tell or hear them. Or, with Rorty's definition of praxis, political activism may replace outworn ideas of truth and liberty. American pragmatism now also becomes politicized and one form of postmodernism. The implications for the loss of actual truth and freedom are enormous, and it is this implication which leads to the likely outcome of the total society. At first, such pragmatism seems to imply a libertarian goal of liberty and to be compatible with a free mind and the least restrictions on behavior that were the value core of our society. But once

this pragmatism became politicized as activism, once the standard for what worked in the world became what ideology was being promoted, this pragmatism transforms to a totalitarian system. Unlike other totalitarian societies, where opposition can still come from classic frameworks, this new total society eliminates any possible basis for disagreement or justification of confrontation.

6. Postmodernism politicizes all social institutions (29, 30). Laclau calls for a continual withering away of the social and, very frightening and important, "the concurrent expansion of the realm of political strategy and argumentation." Botwinick sees the globalization of the postmodern ideology so that its imperial tendencies become acknowledged. There is no denying the conclusion that the use of political power strategies rather than human reason or inductive testing of experience creates a totalitarian society in which critical thinking, free expression, and objective standards of ideas working in the real world are repressed. Only politically-accepted or politically-correct ideas are allowed, and those depend on which groups have raw power in the system. It turns out that Hollinger is wrong and Holton is right about the risks to a free mind that postmodernism entails. The essential intellectual resources of the Enlightenment, as Hollinger phrases it, will be lost in the postmodern world, in fact must be attacked, discredited, made irrelevant, repressed in the postmodern world. What Hollinger overlooked is that postmodernists he names such as Thomas Nagel, Ernest Gellner, Philip Kitcher and Michel Foucault may use terms like "knowledge" and "liberty," but that these terms mean something very different within their political ideology consensus. The terms allow Nagel to be extremely racist in his repression of Europeans and/or Caucasians under the guise of defending affirmative action (31), throwing aside reason and freedom for this class of people because affirmative action suits his political ideology. It allows Thomas Hill to make a strange distinction between civil rights and civil liberties to justify restricting free speech on college campuses and supporting limiting free thought through sensitivity awareness (32). It also explains why patients are not freer in the new health care system than they were in the traditional medical care system, but in fact less free.

7. Postmodernism, with one of its three legs psychoanalysis, therapizes all human institutions (33, 34). The managerial class maintains a therapy standard for management on the surface, while employing global economic strategies as economic first-order decisions. Educational institutions heavily rely on group therapy processes, most mandated, which gradually crowd out the classic content of education. The postmodern focus is on internal emotional and perceptual qualities, having eliminated all but the ground of historicism, as Marxist dialectical materialism was replaced by Neomarxist dialectical

historicism (35). Therapeutic strategies for individual and population manipulation and control, one of the results of Oakeshott's "acceptable" conversations which resemble therapy conversations, promote the ideology in very effective ways, but not very free ways. Playing on self-doubt and psychic weakness, and on group process, the conditioning is subtle and difficult to resist. Resistance becomes an internal problem or flaw of the resister, for there is no other base for it within postmodernism.

The therapy model has implications for our definition of an expert as well. If there is a therapy model rather than a truth/verification model for expertise, then: 1. the expert is evaluated based on sincerity and authenticity, not on knowledge or competence; 2. the expert and his field of expertise is self-creating and without objective standards; 3. the expert's focus will be on textual interpretations, symbol production (public health as public relations), and aesthetic fantasy (see the naming of the personality disorders); and 4. the expert and his field of expertise engages in activism for the political ideology consensus.

8. Postmodernism removes the conceptual underpinnings for human worth. This final implication of postmodernism may seem at first strange to say, since its terminology so emphasizes support for the marginalized and oppressed, for the centrality of human being, for humaneness. But we should not confuse political ideology for the worth and respect of human beings, nor the circularity of the human egocentric predicament for a humane understanding of the human condition. Postmodernism understands the human condition as primarily a political condition, and we should not forget that in the rhetoric. More, it understands the acceptable human condition as the proper political ideological condition, so that worth must be achieved or granted. Human worth is no longer innate, but depends on granting by the political state and on the contingencies of fate which place the person in either an approved (marginal, oppressed) group or an unapproved (oppressor from Eurasia) group. There is no other basis for human worth. Patients, for example, can easily fall into the class of social non-utility, or as units be sacrificed to systems efficiency, with no ground to make a complaint. That ground has been rejected (prematurely and irrationally, but for postmodernism that doesn't matter). The natural rights of man are gone, human rights are gone, the intrinsic worth of an existing person is gone. That follows very simply from the fact that in obtaining political power and keeping it, there is no worth except usefulness toward power.

There is a human crisis point that we are trying to work our way through, and it is a core crisis, fundamental to the human condition, but only if we retain the intellectual tools that allow us to see and understand the human condition – rather than aimlessly act in political praxis. Holton was right and not a zealot in his concern for the Beast, as Hollinger's review wanted to conclude. Carl Sagan, in *The Demon-Haunted World*, was also right to be very worried about the return to the darkness of the mind (36). The possibility of slipping back into Thrasymachus' power view of ethical standards is not only real, but rapidly gaining ground, in particularly bioethics, but in all aspects of our culture. The present danger of losing free thought, reason, and reality is here to see; and if one were to read history carefully for more than ideology but for enlightenment on the ease with which societies slip into the dark, the time is now. Oakeshott may believe that salvation, being saved, has nothing to do with the future, but in postmodernism it has nothing to do with the past or the immediate, either. Political ideology is not salvation for the human condition.

RE-LIGHTING THE ENLIGHTENMENT

Why would it be necessary to intellectually accept postmodernism? It really isn't, unless one accepts the stereotype of a scientifical materialism that primarily existed in the imagination of opponents of the Enlightenment. There is nothing in Thomas Kuhn's work that implies that human knowledge and values are only process, without any valid content. Kuhn shows that central theories (which are paradigms as inductive knowledge grows over time) may either be modified by new theories also based on input from experience, or be placed in a narrower context alongside new core theories, or, if falsified, put aside (37). It would be false to assume that science would be so historically or politically biased that these paradigms were arbitrary human constructs with no basis in reality, or that important aspects of these paradigms did not continue in improved form into a new paradigm. There is no evidence that science has been a cultural story, a human fad in its central method and content. As in any human endeavor, it is difficult to keep bias from somewhat intruding, but the scientific, inductive method is so far the best self-corrective method to avoid this, and for the most part, it has worked well.

Systems Theory supplies an excellent methodology within the scientific method that meets any objection to the possibility of probable knowledge raised by postmodernism. Furthermore, systems theory is central to the scientific enterprise, was developed by scientists, and expanded by scientists, generally from the biosciences but also from physics. Systems theory validates the scientific reduction, again, the most workable method for identifying and understanding the mechanims at play on the next higher level of complexity or perception. This theory is an integration of levels of organization and perception, an inclusive and unifying way of viewing natural experience, and also of viewing ethics (38–41). The HIV infective process, for example, can be seen from a multiorganizational, multifocal point of view, as: molecular cell biology, immune system homeostasis, evolutionary biology, individual behavior and variability, communal systems, and ethology. Reducing to the cellular level allowed us to identify the pathogen and explain what was happening to the person at the physical organism level. Understanding that basic level of the system allowed us to look at the dynamics of the population level such as migration, sexual practices, and breakdown of cultural constraints to appreciate how a lentivirus can operate in human populations.

This inclusive system can ground and develop inductive ethical reasoning as well, as an alternative to the postmodern bioethics. The inductive method of systems theory, applied to ethics, recognizes both the variability of biotic units and the probable generalizations that can be derived (not deduced) from natural experience. Others have shown that the "naturalistic fallacy" is no fallacy, but still balked at deriving "ought" from "is." Yet a scientific reduction such as done by E. O. Wilson and sociobiology is useful (42). Hume would have been pleased to have the biological reduction available to Wilson, and even without it, he basically showed that values could not be deduced, in a formal logic sense, but could be derived from the science of man (43). For example, what we ought to do about AIDS/HIV can be partly generalized as a rule from what we inductively know about its effects on population diversity.

In contrast to the postmodern charge that science aims for completeness and hence is not viable, systems theory is a method and metaphysics of perhaps infinite levels extending in scale up and down our experience of reality. When we talk about objects or matter/energy, we can describe levels of organization or complexity. When we talk about the focus or scale, we can describe levels of perception. Human natural experiences, more or less organized and perceived on larger and smaller scales are called by community ecologists a pattern; by physicists and cardiologists Chaos Theory (44–49). It is true that some scientists, using the old language of the humanities, speak of our "universe" as having a big bang beginning and sounding finite in time and space, but that is more the problem of finding the right words for thoughts. In systems theory (and in a scientific reading of myth likely in the classic world as well), our perceived universe should probably be distinguished from something such as a cosmos where infinite levels could extend. This difficulty with suitable wording does not occur as often in mathematical language, but that is helpful only to mathematicians and some scientists. To the objection that this is a re-characterization of Enlightenment science which has been driven by postmodern legitimate objections and reflects postmodernism, I would point to Liebniz's philosophy of monads, infinite levels of organization and children of the Enlightenment, not postmodernism. I would point to Mandelbrot and his sets which produce fractals, an easily seen representation of systems theory. I would point to Darwin's magnificent theory of evolution, which is systems theory on a paleontological scale. We could even look at the classical scientists and alchemists who thought reality was composed of multiple and often invisible levels of scale, although it was sometimes possible for one level to mediate a connection with another level. That too was part of Enlightenment science.

It is often the case that less-organized or smaller scale is more efficiently maintained over time, but prone to collect into more complex patterns. Although more complex scale is less stable, it can be more flexible and adaptive. A scientific understanding of the less-complex pathogen level and the more-complex human organism level allows us to understand that infectious diseases cannot ever be fully controlled (50). No matter how our political ideology is applied, there is a reality over and above politics that does not rationally allow for a political solution to a cosmic problem.

These multiple levels are connected through feedback mechanisms and this feedback can generally maintain sub-systems unless there are too many critical impacts. Negative feedback mechanisms produce a controlled or oscillating state that can exist over time. These self-correcting mechanisms are not always successful, but it does take more that the flapping of a butterfly's wings to disrupt them. A controlled state can slip into a chaotic state or collapse when overly-challenged, and since these systems are integrated, events at one level can communicate up and down levels. The self-correcting capacity of human DNA for repair, as an example, is only partly successful, and the dynamics of cancer development show that cascading effects on the molecular level can bring down the human system. Given these mechanisms, balance is an important concept.

For science, bioscience, and bioethics, there are no cosmic values in systems. Simpler levels are not "lower" in any value sense, and complex levels are not "higher" (51). Ideology is also a value judgment, and there is no ideology in systems theory. It is human beings who produce in a natural way values or preferences from their human species perspective, as species should do. *It's a wise species that knows what its self-interest is.* The best prospect of achieving pride and reasonableness in that evaluation is not political ideology but scientific systems theory. Bioethics needs to combine the human perspective based on the human natural system with systems realities about ecosystems and galaxy systems to accomplish what is needed and workable for the human species and individual with the least chaotic effects. This is a tall order, but not impossible. It is a worthy order that gives meaning and fullness to human existence. Ideology cannot come close to providing what systems theory can.

The Enlightenment has always been a grander concept and reality than postmodernism.

Systems depend on a controlled balance which is a very complicated endeavor. The internal requirements or needs of various levels often conflict. The feedback mechanisms are imperfect because of time delays in negative feedback. Increasing complexity increases the problem. Often a subsystem does collapse, as the world is profligate with its components. We are not talking about the environmentalism creed of "balance of nature." Such a balance is a myth, as I hope I showed in the first article I had published (52). Chaos, change, disorder, positive entropy are characteristic in systems theory. One who expects the garden of Epicurus had best resign himself to the chaos of Tiamat.

Public policy ethics, or public health ethics, should both be concerned with systems balance rather than ideology. Postmodernism has no concept of balance, no methodology to deal with one of the fundamental questions of science and human existence. Political balance of power is not the answer to systems controlled states. Power politics is not the answer. Stories or myths about a non-existing balance of nature cannot be helpful Conversations remain conversations, nothing more. Only scientific systems theory can supply the understanding we need to work toward maintaining controlled-state balances that are important to us as a human species.

There is a danger that as our science becomes corrupted with ideology, and that can occur despite the self-correcting of the scientific method when there is sufficient cultural force to generate it, we will lose what we need to describe and intervene in systems balance. Chasing the false ideology of the wonderful balance of nature (only disturbed by one species, the human one), we could lose the chance to maintain those controlled-state balances that are very important to our self-interest. Scientific systems balance is a key concept in managing the AIDS/HIV pandemic; in looking at infectious diseases that are emerging, resurging and resistant; in constructing medical care delivery systems; in reading the human genome; in reconciling the abortion positions; in defining biological variability for disease classification; in reproductive medicine research; and in decision-making about terminal care.

In Part II, I will construct an argument that it is not postmodern bioethics that will be a pragmatic and rational tool for resolving medical care issues, but rather the evolving science that re-flowered in the Enlightenment and, despite objections from postmodernism, continues to increase our understanding of that perceptual slice of the cosmos we call the universe. It is in our species interest to continue to carry that light of discovery and verification. Bioethics can be a true interdiscipline.

REFERENCES

1. Sulloway, F. J. (1979). *Freud, Biologist of the Mind: Beyond the Psychoanalytic Legend*. New York: Basic Books.
2. Botwinick, A. (1993). *Postmodernism and Democratic Theory*. Philadelphia: Temple University Press.
3. Bohm, D., & Hiley, B. J. (1993). *The Undivided Universe*. London: Routledge.
4. Botwinick, A. (1993). *Postmodernism and Democratic Theory*. Philadelphia: Temple University Press.
5. Botwinick, A. (1993). *Postmodernism and Democratic Theory*. Philadelphia: Temple University Press.
6. Connor, S. (1989). *Postmodernist Culture: An Introduction to Theories of the Contemporary*. Cambridge, England: Blackwell.
7. Paris, J. D., Cronte, R. K., & Reardon, F. (1990). Physicians' refusal of requested treatment: The case of Baby L. *New England Journal of Medicine*, *322*(14), 1012–1014.
8. Orentlicher, D. (1999). Medical ethics and the law. In: R. B. Edwards & E. E. Bittar (Eds), *Advances in Bioethics*, pp. 101–112.
9. Veatch, R. M. (1993). The impending collapse of the whole-brain definition of death. *Hastings Center Report*, *23*, 18–24.
10. Jecker, N. S., & Schneiderman, L. (1992). Futility and rationing. *American Journal of Medicine*, *92*, 189–196.
11. Botwinick, A. (1993). *Postmodernism and Democratic Theory*. Philadelphia: Temple University Press.
12. Connor, S. (1989). *Postmodernist Culture: An Introduction to Theories of the Contemporary*. Cambridge, England: Blackwell.
13. (Mitchell, E. J. T. Ed.) (1983). *The Politics of Interpretation*. Chicago: University of Chicago Press.
14. Oakeshott, M. (1991). *Rationalism in Politics and Other Essays*. Indianapolis: Liberty Press.
15. Callahan, D. (1987). *Setting Limits: Medical Goals in an Aging Society*. New York: Simon and Schuster.
16. Wilfore, B. S., & Fost, N. (1990). The cystic fibrosis gene: Medical and social implications of heterozygote detection. *JAMA*, *263*(20), 2777–2783.
17. Englehardt, Jr., H. T., & Wildes, K. W. (1994). The emergence of secular bioethics. In: E. Bittar & N. Bittar (Eds), *Principles of Medical Biology*, Vol 1A (pp. 1–15). Stamford, CT: JAI Press.
18. Bayertz, K. (1994). *Consensus*. Kluwer Academic Publishers, Dordrecht, The Netherlands.
19. Hollinger, D. A. (1994). Embattled values. *Science*, *263*, 252.
20. Oakeshott, M. (1991). *Rationalism in Politics and Other Essays*. Indianapolis: Liberty Press.
21. Botwinick, A. (1993). *Postmodernism and Democratic Theory*. Philadelphia: Temple University Press.
22. Connor, S. (1989). *Postmodernist Culture: An Introduction to Theories of the Contemporary*. Cambridge, England: Blackwell.
23. Botwinick, A. (1993). *Postmodernism and Democratic Theory*. Philadelphia: Temple University Press.
24. Connor, S. (1989). *Postmodernist Culture: An Introduction to Theories of the Contemporary*. Cambridge, England: Blackwell.

25. Botwinick, A. (1993). *Postmodernism and Democratic Theory.* Philadelphia: Temple University Press.
26. Connor, S. (1989). *Postmodernist Culture: An Introduction to Theories of the Contemporary.* Cambridge, England: Blackwell.
27. Botwinick, A. (1993). *Postmodernism and Democratic Theory.* Philadelphia: Temple University Press.
28. Rorty, R. (1989). *Contingency, Irony, and Solidarity.* Cambridge, England: Cambridge University Press.
29. Botwinick, A. (1993). *Postmodernism and Democratic Theory.* Philadelphia: Temple University Press.
30. Connor S. (1989). *Postmodernist Culture: An Introduction to Theories of the Contemporary.* Cambridge, England: Blackwell.
31. Nagel, T. (1995). Equal treatment and compensatory discrimination. In: J. Feinberg & H. Gross (Eds), *Philosophy of Law* (pp. 429–435). New York: Wadsworth Publishing.
32. Hill, Jr., T. (1995). The message of affirmative action. In: J. Feinberg & H. Gross (Eds), *Philosophy of Law* (pp. 436–450). New York: Wadsworth Publishing.
33. Botwinick, A. (1993). *Postmodernism and Democratic Theory.* Philadelphia: Temple University Press.
34. Yankelovich, D., & Barrett, W. (1971). *Ego and Instinct: The Psychoanalytic View of Human Nature – Revised.* New York: Vintage Books.
35. Connor, S. (1989). *Postmodernist Culture: An Introduction to Theories of the Contemporary.* Cambridge, England: Blackwell.
36. Sagan, C. (1995). *The Demon-Haunted World: Science as a Candle in the Dark.* New York: Random House.
37. Kuhn, T. S. (1990). Structure of Scientific Revolutions, 2nd ed. Chicago: University of Chicago Press.
38. Von Bertalanffy, L. (1968). *General System Theory: Foundations, Development, Application.* New York: George Braziller.
39. Gray, W., & Tizzo, N. D. (1973). *Unity through Diversity*, Vols. I and II. New York: Gordon and Breach.
40. Clements, C. D.(1985). Therefore choose life: Reconciling medical and environmental bioethics. *Perspectives in Biology and Medicine*, *28*(3), 407–425.
41. Landau-Stanton, J., & Clements, C. D. (1993). *AIDS, Health and Mental Health: A Primary Sourcebook.* New York: Brunner/Mazel.
42. Wilson, E. O. (1978). *On Human Nature*. Cambridge, MA: Harvard University Press.
43. Hume, D. A. (1888). *A Treatise of Human Nature: Being an Attempt to Introduce the Experimental Method of Reasoning into Moral Subjects*, Selby-Bigge, L. A. (Ed.). Oxford: Clarendon Press.
44. Pickett, S. T. A., & McDonnell, M. J. (1989). Changing perspective in community dynamics: A theory of successional forces. *TREE*, *4*(8), 241–245.
45. Pickett, S. T. A., Kilosa, J., Armesto, J. J., & Collins, S. L. (1989). The ecological concept of disturbance and its expression in various hierarchical levels. *Oikos*, *54*, 129–136.
46. Pattee, H. H. (1973). *Hierarchy Theory: The Challenge of Complex Systems.* New York: George Braziller.
47. O'Neill, R. V., DeAngelis, D. L., Waide, J. B., & Allen, T. F. H. (1986). *A Hierarchy Concept of Ecosystem.* Princeton, NJ: Princeton University Press.
48. Krasner, S. (1990). *The Ubiquity of Chaos.* American Association for the Advancement of Science, Waldorf, MD.

49. Kolasa, J., & Pickett, S. T. A. (1989). Ecological systems and the concept of biological organization. *Proceedings National Academy of Science, U.S.A.*, *86*, 8837–8841.
50. Clements, C. D. (1989). Biology, man, and culture: A unified science based on hierarchy levels. *Perspectives in Biology and Medicine*, *33*(1), 70–85.
51. Clements, C. D. (1983). The lesion and the function: Setting up the reductionist problem. *Perspectives in Biology and Medicine*, *26*(3), 433–440.
52. Clements, C. D. (1976). Stasis: The unnatural value. *Ethics*, *86*(2), 136–144.

PART II

SPECIFIC ISSUES IN BIOETHICS: ANALYSIS AND VALIDATION OF THE ARGUMENT

4. HUMAN REPRODUCTIVE MEDICINE: FREEDOM OR REGULATION

THE BASIC HUMAN VALUES OF FERTILITY AND REPRODUCTION

An interesting thing happens when bioethics is reduced to political process and ideology. The full humanity of individuals then also becomes reduced to that same process and ideology, and we lose what makes us human beings. There is nothing sadder, stranger, or less human than the new medicine that academic bioethicists, politicians, and lawyers have tried to create. A philosophic understanding of the human condition is rapidly being lost in that institution of society where it was once most understood, medicine. "Health care" rather than "medical care" is changing into a strange combination of college classroom lectures, legal arguments, political ideology, and social manipulation of lifestyles. The new postmodern bioethics has ironically produced a cold, technical, bureaucratic, and basically non-human medical model while criticizing the previous clinical care as based on a medical rather than wholistic or human model. It has replaced that scientically and humanly aggressive portrait of a physician in *Ben Casey*, with its opening lines centered on the deep issues of being human (man-woman, birth-death, infinity). Instead, the public is given a MASH surgeon spouting political ideology while fighting the non-politically-correct military establishment and politicians, an ER physician with all the ideologically-correct points of view briefly interacting with patients but more concerned about private angst, and a barely-stable cardiac surgeon who sings and decides to become human by being a clown. Popular culture can say a great deal about the medical ethics discontinuity.

This loss of humanity has happened in many specialties and clinical situations in medicine, for example, in the way death and dying is now treated as a college course on information about the process, or a politicolegal consultation about rights and treatment refusal, or a beauty contest about superficial "dignity," or a programmed therapy session about stages. Nowhere, however, is the loss more poignant or central to the fundamental values of our human species than in reproductive medicine.

There is a studied disregard for the importance of the genetic message and the physical evolution of that message from the material of stars to the changing biodiversity of our planet. The programmed striving to pass on that message, through sexual reproduction in humans, is not celebrated or valued as it once was in all our cultural memories. Fertility is no longer a goddess. Postmodern culture and bioethics does not have the capacity to see the wonder of passing on the genetic message. Birth has become a political act of choosing a midwife over a male physician, enduring pain and the risk to the baby of a long labor in order to avoid "unnatural" drugs and technologies, opting for increased-risk home delivery to wrest power over females from male institutions like hospitals – or choosing not to burden the planet with another member of the valueless human species. We need only read E.O Wilson's wonderful prose about genetic life to realize how fundamental, how profound this genetic nature of man is and what fully is at stake: "This is the assembly of life that took a billion years to evolve. It has eaten the storms – folded them into its genes – and created the world that created us. It holds the world steady" (1).

A fair look at Wilson's sociobiology can allow us to appreciate the importance of the genetic blueprint, as Matt Ridley calls it in his book, "the autobiography of a species in 23 chapters" (2). Because the neoleft ideology of much of bioethics rejects genetics as a proper political stance, it downplays and negatively portrays one of the most astounding leaps in knowledge that our human species has made. Postmodern bioethics refuses to understand why fertility and reproducing should be so important to individuals, barely giving their wishes/choices lipservice before trying to curtail and denigrate them. It portrays such people as under a coercive pressure than makes them victims or questions their capacity to consent. It suggests such focused wishes may be selfish. It implies that, given the risks of multiple births or reproductive technologies, such choices might be unethical child abuse of the future delivered babies, one of the few times that postmodern bioethics considers the fetus to have a moral standing. It believes and maintains that such a valuation of genes being passed on to the next generation is mistaken, and that heroic attempts to do just that should be viewed with the suspicion that they are unethical, a prima facie devaluing of fertility treatment and reproductive medicine.

MULTIPLE BIRTHS AND THE BIOETHICS RESPONSE

Nkem Chukwu, a Nigerian-American who gave birth to octuplets in December of 1998, provoked the same response from bioethicists that Bobbi McCaughey did when she gave birth to septuplets in November of 1997. In both these cases, the multiple births were used as an opptunity to gain control over one of the few remaining medical specialties that is still primarily controlled by clinical experts.

Reproductive medicine ironically benefitted from the ideological attempt by bioethics to control it through the banning of federal funds for the research needed to develop and advance the field. In vitro fertilization (IVF) was the first procedure to be cut off from national funding in the U.S., with the goal that such research and development censorship would severely limit an expensive medical procedure and draw funding toward the more ideologically-acceptable field of population control. However, the needs of infertile couples and individuals were so primary that this was sufficient to create a privately-funded system. Because the system was privately funded, reproductive medicine avoided the centralized control that bioethics was supporting. Reproductive medicine managed to escape bureaucratic control by non-physicians and remained under the leadership of medical experts.

This control by professionals of the professional institution was unacceptable to non-medical managers and humanities experts as well as lawyers specializing in medical law. Bioethicists, following the births of both the Chukwu and McCaughey babies, made the media rounds in much the same way the political spin doctors make the rounds, and the comments were all remarkably alike, falling into a predictable list of main concepts, or, as a politician would say, "talking points."

First, the problem of multiple births was greatly exaggerated. The public was not given figures to place the question in the real world context. For example, in 1995 there were only 11,631 IVF babies born in the U.S., so the specialty of reproductive medicine does not have a large impact on society in terms of people using the technology. In 1994, the births of triplets, quadruplets and quintuplets totalled 4,594, a modest rise from 1,005 twenty years ago. But we need to separate out twins, triplets, and larger multiples, and the bioethicists did not do that. Instead, they took the health risks for all multiple births,which skews the figures. Prematurity was also discussed globally rather than broken down into the risks for weeks of gestation. There are very different risks for different weeks. Even the social and psychological risks were overstated. A spokeswoman for the National Organization of Mothers of Twins wondered how parents could have enough clothes and toys for four to eight babies, forgetting that church, extended family, and community are generally present for support, and that material goods are not the only criterion of happy childhood. A psychologist doubted the parents could give the necessary attention to all the babies, thus depriving them of the care and love they needed and ruining them for life, again forgetting that there are such things as extended families, church groups, and often a network of close friends. These last objections indicate a materialistic value hidden in postmodern bioethics, which reflects a political aim toward redistribution of economic resources and its implication that monetary or material values are the road to human good. I will discuss further the distortion of data to fit certain ideological assumptions and the value bias in bioethics when I look at the McCaughey case.

In both cases, selective reduction was actively pushed by bioethicists who commented. A New York State Health Department special panel has urged physicians to discuss with their patients aborting some fetuses in high-multiple pregnancies, reflecting the close connection between bioethics and public policy officials. Although bioethics has insisted that genetics counsellors be non-directive with patients (3), that does not apply to multiple births, even though no good reason is given for the difference. The standard of non-directiveness was easily cast aside for, I can only conclude, ideological reasons. It was seen as not only ethical, but recommended as an obligation that the medical community encourage selective reduction. In both the septuplet and octuplet cases, bioethicists criticized the parents for not considering selective reduction and jeopardizing the well-being of the babies. This moral standing for seven or eight fetuses does not extend, however, to single fetuses that are aborted routinely based on the woman's choice. One professor of bioethics even argued that women who would refuse selective reduction (this is actually abortion of some of the multiple fetuses) for religious or other reasons should not be allowed to have fertility treatment. He argued that opposition to aborting some of the fetuses in a high-multiple pregnancy was a disqualifying factor for fertility treatment. This coercion to have an abortion was apparently a different sort of right to choice than coercion against having an abortion. The rational contradiction can only be explained by recalling the basic irrationality and ideology of postmodernism. No one addressed how such a requirement was to be enforced. What would the state do with women who had lied or changed their minds? Would selective reduction be forcibly done? Would the entire pregnancy be aborted? Would treatment be refused during the pregnancy? Would there be financial fines? Would such women be blacklisted at treatment facilities from then on? Would U.S. Supreme Court decisions against any substantial coercion of a woman's right to choose abortion (4) also apply to a woman's right to choose not to abort?

Another point was the cost of multiple births, straightforwardly presented as a reason for controlling reproductive medicine, where twenty years ago, that goal would have been concealed or not admitted. The medical bills for such pregnancies were described as enormous and falling on everyone, picturing such couples as a drain on social resources rather than as contributing members to social resources. These "enormous" expenses were then described as avoidable and the individuals incurring them for all of us as irresponsible. Another professor of bioethics even suggested that insurance companies may either want to increase their premiums or deny coverage for "aggressive" fertility treatments. He thought it would be a perfectly reasonable and legitimate thing for insurance companies to do. The implication was that such individuals will raise all our insurance rates, but there was more. Coverage of treatment for the babies born might also be denied, a powerful economic weapon against couples using reproductive medicine.

Finally, the bioethicists recommended more regulation of reproductive medicine as the solution to its many hypothetical problems. Bioethics' political ideology trends toward regulation and centralization as the answers to most medical issues. But clinics already have rules concerning quality of care: legal requirements of medical confidentiality apply, infectious disease control and universal precautions are generally state-mandated, licensing and credentialing must meet state requirements, and clinical standards must be met. With these regulations already in place, what regulations are being so vigorously promoted? Such new regulations would perhaps require aborting some of the multiple fetuses ("selective reduction" would be known in inductive logic as a "weasel word"), downsizing fertility treatments available, and restricting research and development in reproductive medicine. These regulations would likely aim at removing medical experts as the decision-maker, and replacing them with bioethicists who would bring their ideology and economic cost-efficiency to the system. This may be the real reason for the extensive media coverage of high multiple births, using them as trial balloons to complete the politicizing of the medical care institution.

But there is still the reality of the two cases. The McCaugheys had experienced a fertility problem but were successful in conceiving and having a daughter using one of the standard fertility drugs available. Since they wanted a large family, they decided to again use the drug, at the same low dose that was successful before. There is the risk with these drugs that multiple pregnancies will result, but in 80% of such multiples, it is twins, with low health risks for all. The risk of quadruplet or higher pregnancies is less than 5%, but with such multiples the pregnancies may end before 30 weeks, when there is greater risk that the babies either will not survive or will be damaged. The actual quadruplet births in the U.S. for 1995 were 365; for quintuplets, 57. In other counseling, the experts would have been taught by bioethics to discuss both the 5% risk of a bad outcome and the 95% risk of a good outcome in taking fertility drugs, and the counselors would also have been told to maintain a neutral position. In my experience with our Prenatal Detection Program counseling, bioethics insisted on such neutrality even for high risk genetic diseases such as Huntington's (50%) or Cystic fibrosis (25%) (5). But again, if we realize the political/ideological nature of bioethics, this makes some sense. Genetics is considered to be ethically questionable, associated with eugenics which is assumed unethical, and implying biological determinism which is unacceptable to the neoleft. So bioethics demands complete freedom of choice for reproduction when the context is genetic, no matter how severe the genetic disease. On the other hand, terminating pregnancies and reducing population are ideologically compatible, but fertility is suspect.

In analyzing the McCaughey case for the media, some bioethicists were not happy that the outcome had been relatively good. One argued that a good outcome makes bad policy, law, and ethical discussion. Unless the ethical discussion can be driven by scandals and disasters (here we should translate as "worst-case trial balloons"), unless all seven babies had been born dead or defective, the bioethicists could not easily stop and ask how it could have happened, how it could have been allowed to happen. Unfortunately for the bioethicists, Bobbi McCaughey made it past 31 weeks gestation and delivered seven live babies. All eventually went home, where two may have some serious residual effects but all are cared for and stable. The bioethicist was wrong. Good ethical analysis is based neither on happy outcomes nor scandals and disasters in trial balloon cases. It is based on a careful contextual look at the full picture and a viable ethical theory of content rather than process. However, the bioethicist was also right in one sense, or honest. If bioethics is defined as a political public-relations tool, then it *is* based on trial balloon cases of scandal and disaster. We might even conclude, except for a base of ideological assumptions, it is based on nothing more than scandals and disasters.

There is a serious and unanswered question about applying social/political policy to couples' reproductive needs and interests. Is centralized control of such a basic human value an ethical position? In other words, why should such reproductively-restrictive regulation be brought into multiple birth cases at all? Could any reproductive health policy be ethical? If we think that this is a major issue in human freedom, and that bioethics recognizes it as such, we would again be mistaken. Another bioethicist and lawyer considers fertility treatment to be a "market good." Seeing reproduction as a market good, he can advocate economic gatekeeping so that some individuals will be restricted based on age, number of attempts, risk of multiple births, and just distribution of resources. Bioethics of this sort reflects the ideological consensus and easily replaces autonomy with coercion and a totalitarian set of guidelines. This bioethicist/lawyer would require that women agree in advance not to carry more than three or four fetuses. If they refused to agree, that would be proof they were not ethical enough to protect their child.

If the technique were IVF, such a requirement could be enforced by not implanting more than three or four fertilized embryos. But the more common, routine technique of fertility drugs cannot be so easily enforced. Should the couple have to abstain from sex when the number of ripened eggs is too high, even if this decreases their chances of ever having a baby? How could abstention be enforced? If all else fails, can the pregnant woman be required to abort some of the fetuses?

These serious ethical and political questions have not stopped bioethicists from forging ahead on health policy for reproductive medicine. A lawyer who is a health policy expert wants the U.S. to have a national policy on multiple births, a totalitarian approach not seen for what it is. She justifies the approach by distorting statistics that 90% of multiple births are born premature, where only 10% of single babies are. The actual weeks of prematurity are not cited, nor are variables among population groups, and prematurity is considered as all babies born less than 37 weeks gestation. Her figures ignore more precise statistics that triplets born at 33 weeks have a 96% survival rate, ignore all twinning rates, and do not address severity of morbidity for varying prematurity (6). Based on such incomplete statistics, bioethics is willing to make reproductive decisions a health policy, removing the decision from the individuals and placing it with bureaucracies of centralized government. I can't think of a better definition of the total society.

Reproductive medicine is described by bioethics as an industry estimated to gross $4 billion a year, with the key dyslogistic terms "industry" and "$4 billion." This plays on class warfare and current ideological dislike of anything involving business or profit. Bioethics wants to bring this system under the managed care system so it can be downsized (rationed, censored, utilization controlled).

The McCaughey case also displeased bioethics because the couple had deep religious convictions and community, and refused to consider selective reduction. They did not agree with the ethical utilitarian position of sacrificing some to save others, and instead took the ethical position that either all would survive or none would, both arguable in philosophy. Being opposed to abortion in general, they did not make the politically-correct reproductive choice, since bioethics does not really consider all reproductive choices as equal. The couple seemed too close to religious conviction rather than secular ethics, and it is permissable under postmodernism to discriminate against religious beliefs that are not part of the ideological consensus. I am comfortable making that statement because I am a nontheist with no religious affiliation, who nevertheless feels the concept of free thought is a basic human value for all, and that arguments should be objective and rational, not political and manipulative. For example, the attempt by the International Humanist and Ethical Union to have the Vatican state removed from membership in the United Nations because of its influence on European members is a political strategy of suppression of free thought and expression. Since such thought control violates what were once the ethical values grounding humanism and science, I can only conclude that the IHEU is another institution politicized with the ideological agenda of postmodernism (7).

At the December birth of the Chukwu babies, even with one of the eight dying after being born earlier, we should have been celebrating as a culture the awesomeness of life, the evolving patterns of genes that create the world and us. Instead, the cultural changes since the 1960s and 1970s have devalued fertility in favor of managed, efficient, static, and eventually sterile political behavior. We have lost the history of the ancient kingdoms of the Middle East who celebrated messy, unrestrained, powerful, vitalistic fetility. We have forgotten the fragmented strivers of the Medieval period who should be celebrated for slogging and courageous human reproduction in the face of all the calamities of the English Book of Common Prayers. Those societies didn't have to worry about overpopulation, of course, but there is something fundamentally human that is lost when fetility becomes a negative, when children are seen as a problem rather than a hope. Spengler saw this as one of the first symptoms of the cyclical change from a civilization on the rise to one in decline (8). If that is only partly accurate, we have still come far down the road, because Bobbi McCaughey and Nkem Chukwu are culturally presented as anachronisms, women from another time (9).

OLD WOMEN AND NEWBORNS

Not only women who carry multiple pregnancies to term are targets of postmodern bioethics. The postmodern ideology of bioethics is also critical of older women who use reproductive technology to have a baby. This issue is another example of the inconsistency and contradictions that are an integral part of bioethics. It should be expected, since postmodern bioethics is based, as I hope Part I demonstrated, on a philosophy that denies the importance or relevance of logic (reason) and verifiability (reality). Instead, clinical decisions such as assisting a post-menopausal woman to have a baby are now supposed to be driven by political ideology and enforced through power strategies. The proliferation of contradictory ethical positions, the arbitrariness of bioethics, is a serious concern.

There have been a few cases of women in their sixties who have used reproductive medicine technologies to have a baby, and the bioethics consensus has raised a number of objections.

Old women will die before the children are in high school. There is an increased probability that, statistically, some might, although chronological age is not the same as biological age. The probability, over a large class, is that less young women will die before their children are in high school. But for the single case, we only have gambling figures. The older women who now have a child might live to see the child marry, and the younger women who have children might die in car accidents or from an unexpected disease. Is the risk so high, so severe, so removed from the individual's control that it is an ethically-unacceptable risk? Given the increase in life expectancy, these women may easily reach 78 or 80 years of age. Their child may be 16, 18 or even 20 before death takes the mother. This will be a loss, as the loss of any parent is through death or divorce, but it is a bit strange to ignore the context of general human experience with loss or of single-parenting because of divorce which is more and more common. The bioethical goal cannot be to protect human beings from all the risks of the human condition, since such a totalitarian agenda must be unethical. However, in the service of political ideology, bioethics has trended toward such a protective, freedom-restricting, risk-restricting goal, and we would be mistaken to assume bioethics is in disagreement with such a total society. It is rather the opposite that is likely to be true.

The second important question in this analysis is to ask what entity is being protected from the risk of loss? It is not the existing woman and man who wish a pregnancy and child. They have already chosen to take the risk, have made an assessment and evaluation that is adequate given the context of normal human risk and loss. No, the entity to be protected is the "future" child, the still non-existent child that somehow has more moral standing than a six-month fetus whose mother wishes an abortion for any reason. It is the unconceived child that is being protected by bioethics, protected by not being allowed to come into existence. Although bioethics will not limit the reproduction of a woman carrying the Huntington's gene, where the severe risk to such a future child is 50%, it is seeking to limit the reproduction of an older woman because of the risk that her teenage child will suffer the loss of the mother. It is only with the irrationality and meaninglessness of postmodern conversations that such a line of argument could even be constructed. That it is an invalid and ridiculous attempt at reasoning is not even considered by current bioethics, and there lies a great problem for contemporary society as well as older women.

Old women, bioethics argues, have greater health risks in pregnancy than women in their twenties and thirties. So far, the medical histories of older women's pregnancies have not supported that claim, but the sample is small. It is a high-risk pregnancy by definition, but perhaps not by reality. The reproductive health of the older woman is probably the strongest determinent of outcome, and that can be partially assessed. It is probable that these pregnancies will be a higher risk, but not an extreme risk, than the general class of pregnancies, but again, the question of how to apply such statistical figures to the single case (The Single Case Problem) is not so easily resolved. We should also not forget a full context. Teenage pregnant women also have slightly higher risks than the general class. Should a happily-married, healthy fifteen-year-old who wishes to start a family but requires the intervention of reproductive medicine be denied that medical technology? For state regulatory agencies to make such decisions for emancipated minors is disturbing in terms of a trend to a totalitarian society. Postmodern bioethics does not support any "paternalistic" or family constraints on emancipated minors for a variety of decisions: refusing life-saving treatment, having an abortion, receiving routine medical care, choosing a residence, deciding on psychiatric care. If teens are allowed to take health risks of those sorts, why should older women be refused the choice of taking a health risk with a pregnancy?

How risk is evaluated is a thorny ethical question that I will address in the next chapter. For now, I want to argue that prudence is not always the reasoned ethical choice. Some risk taking is ethical and a needed component of our human nature, our human condition. It is one of the basic values of our existence that needs reaffirmation in current culture. Freud understood what a central part of human existence risk taking might be, I think, when he used the Dark Ages Hanseatic League motto in his writing: "It is necessary to sail; it is not necessary to survive." That is a profound description of our human behavioral transaction with the natural environment.

Bioethics has also objected to old women having babies because it is assumed that old women do not have the energy to raise children. The postulated entity, the future child, will somehow suffer child abuse, improper parenting, because the older woman will be exhausted. Again, biological health rather than chronological age is more important here in making a risk assessment, and individual variability is paramount. Some older women have spent months paddling in the Everglades, have climbed mountains, have gardened over acres. Even the average older woman can find the energy to raise children, as shown in the African-American communities where grandmothers are often the primary care-giver, the substitute mother. I can recall interviewing a severely-schizophrenic young woman who had been hospitalized. Her baby boy had been taken away from her because all she fed him was a bowl of milk. Everyone assumed this was another bit of evidence of her psychiatric pathology. Still, there was something very maternal in her wishes, and I began to ask her about her childhood. She told me she had been raised by her grandmother in a small cabin in the South, where money was hard to come by. Still, she said, her grandmother would always make sure that there was a bowl of milk for her at the table, and her eyes filled with love as she described this woman. Actually, she was trying to be as good a mother to her little boy as her grandmother had been a mother to her. It was a case I never forgot.

For centuries, in family crises, grandmothers have been stepping in to do the mothering, no matter what the age (10). What needs to be done will be done. It may not meet some artificial standard of parenting prescribed by health experts, but it has generally supplied love, comfort, stability, a home and a family. It was done by old women. Is there any reason why an old woman could not choose mothering a child of her own? We need to ask on what basis bioethics would use the power of the state to prevent such women from having babies with reproductive medicine technologies.

Bioethics argues that old women are violating some rule of nature that makes menopause the end of reproduction. On this feminists are oddly silent, and I will have more to say about this silence later. What such arguments deliberately overlook is that all medicine violates some "rule" of nature, meaning medicine does not stand passively and let nature takes its course, or allow the continuation of a certain process when it could be stopped. It also fails to establish the definition of what such rules of nature are, because postmodern bioethics does not have the scholarship of natural law behind its ethical (political) analyses. One might attempt this argument using natural law theory, although I think it would ultimately fail. But postmodern bioethics does not have such a theoretical underpinning, and it is inconsistent to call it forth only for a convenient ideological goal.

Medicine is as interventionist as human nature is, and it does intervene to prevent dying from infectious diseases. It intervenes in trauma situations and always has. It can now intervene in some genetic defects, and it is beginning to build some success with chronic diseases, if not with mortality, at least with morbidity. It makes no sense to advocate, in one specific area of reproduction, that the interventions of medicine be curtailed, simply on the basis that it is wrong to intervene in "natural rules," whatever those might be. To attempt such an argument is to illustrate the weakness of natural law arguments about contraception, some abortions, IVF, donor eggs, donor sperm, fertility medications, and now, pregnancy in older women. Many more premises are needed to make that argument, and some of those premises are suspect. However, natural law theory at least does provide some of those premises, whereas postmodern bioethics is empty of them.

At the base of the bioethics objection is the assumption rather than argument that older women who want to bring new life into the world should not be allowed to make that choice or decision. Reproductive choice is not allowed for older women. Or, if considered at all, unlike the reproductive choice for abortion which requires no reasons whatsoever, reproductive choice of older women must have reasons even beyond what those women can give: guarantees for the future, high standards of superior parenting, and proof that natural rules have not been violated.

At the same time, bioethics does give older women who have a terminal disease, or who simply want to refuse treatment for whatever personal reason, the right to make that decision and die. Many bioethicists want to extend that right to include assistance in suicide or direct termination by physicians. As long as an older woman wants to take life out of the world, bioethics supports and facilitates that decision, with no reasons required.

Why has there been such criticism of older women's pregnancies by both male and female bioethicists, when such pregnancies are life-affirming in the most stubborn of ways? Why have these small number of cases generated any comment at all (11)?

There are psychological reasons, though I do not think they are the most important ones. There is a social bias against old women that leads perhaps to revulsion at the idea that they could become mothers. This bias is apparent in the sexual area, where it is denied that old women can be sexually interested or interesting. In the aesthetic realm, older women are seen as not attractive enough to be mothers, not fitting our picture of the beautiful people. Young women look good holding babies, Madonnas. Older women do not make as attractive a picture. The "beautiful" people are healthy people who will never grow old and die, and who can generate babies without the reminder of mortality. I doubt, though, that bioethics has that full or profound a view of life, since politics does not have that encompassing a picture of humanity. I would argue the reasons for objecting to older women having babies are much more ideological, political and economic.

The economic factor is tied closely to the political assumptions of postmodernism. It represents trust that a centralized control will deliver more efficient and "just" prioritizing of resources. It constructs a bureaucratic system of appointed experts not responsible to the polis or the republic who will make economic first-order decisions based on their own consensus and the unexamined values of the official class (12). It contains a neoleft assumption about egalitarianism in distributing health care resources, implying that none can have what all cannot have, but actually effecting a rationing system that lowers quality and access for all, even in contexts where there is no real scarcity. Finally, it assumes that a planned, preventive intervention is feasible, acceptable and superior to interventions for disease and trauma. In such a system, reproductive medicine is a free market anarchy that will not be tolerated. The argument is that reproductive technologies such as IVF are expensive and need to be restricted, and that they are used by a privileged few to the deteriment of broad social and preventive programs for health.

Even though third-party payers may not reimburse for much of reproductive medicine, so that health dollars do not come directly out of the preplanned and regulated health care system, bioethicists will argue that society still pays the cost indirectly, even if the system is a private system. The research and development done privately in reproductive medicine, and its utilization of general world resources, the argument goes, uses up the postulated resources available in the global system, and would be better channeled to other health issues that emphasize prevention. This argument is made in the context of strong trends in current bioethics to limit and censor medical technology, partially based on 70s studies by The Club of Rome (13). Those computer models have not proved accurate, but the ideology remains that technology is assumed to be unethical unless regulated into "appropriate" technology, so that less is ethically better, the "small is beautiful" assumption (14). No rational cost/benefit analysis has actually been done concerning technological research and development, but this ethical misperception is nicely captured in a favorite bioethics phrase: "The Technological Fix." The evolving product of the human learning program, technological application and breakthrough, is negatively evaluated as a scientism, a human assault on the natural system, and false pride. In reality, this argument collapses, but we are talking about ideology, not a reasoned historical analysis of human technology and its risk/benefit.

The political assumptions make reproductive technology unacceptable to bioethics. Its neomarxist position conflates technology with eugenics and Frankenstein Syndromes such as cloning and genetic cleansing. An evolutionary approach such as sociobiology, or the application of genetic engineering and biotechnology are presumed equivalent to fascism. The bias is even deeper than that. Assisted reproduction is recognized as a biological description of human behavior, an acknowledgment of the scientific reduction, a tool for genetic engineering which forces recognition that genes are determinants and that there is a human nature which limits human probabilities through the arrangement of human DNA. This is contrary to neoleft doctrine that human beings are completely pliable and that social systems, not our biology, shape what we will become. Political praxis can change and construct social systems, it is believed, so politics can create human beings in its own image. The new microbiology, the human genome project, and reproductive medicine all threaten such a political ideology. Since this ideology is the base of current bioethics and its politicizing of the medical care institution, the biological synthesis that has opened up our understanding of the human biological system and the constraints of reality must be rejected. Reproductive medicine is one application of that molecular biology synthesis that Ernst Mayr has so well described (15).

It is superficially puzzling that feminists have not come to the defense of older women, but it is understandable on a deep analysis of the postmodern ideology of which current feminism is an integral part. It is neoleft politically and reflects the assumptions of that ideology, along with the preference for those defined as "victims." The goals it sets for young women have little to do with heterosexual relationships, mating or families. The ideal new woman looks very much like an adolescent boy, with short hair, anorexia-produced lack of mature secondary sexual characteristics, and power-dressing to become the dominant figure in society (actually, the oppressor). A woman is expected to have a high-power career that competes with males and overcomes them. She is supposed to be capable of having it all, but if that proves difficult, then reproduction should be postponed until she has achieved her career heights. Generating children is not a major value, and those children are quickly plugged into the centralized plan of political institutions, daycares and mandated preschools. Those who interact with the ideal woman are supposed to deny her sexual biological nature, at the risk of lawsuits.

It is no surprise, then, that current feminists support abortion and the right of old women to die, but are uncomfortable with the choice to continue a pregnancy or for old women to choose to use reproductive medicine to become pregnant. That is not the feminist political ideology. The response to old women having babies is therefore ambiguous and internally contradictory, but we have seen that this is not a problem for postmodernism which is internally contradictory by choice. It is a problem, however, for women, and particularly old women, who wish to express their full nature and not become dehumanized.

CLONING, SEXUALLY REPRODUCING, AND READING THE HUMAN GENOME

Speaking before the 4,000 members of the Transplantation Society, Pope John Paul II condemned the cloning of human cells to make life-saving transplants. Although the pope praised organ donation as an expression of brotherly love, he criticized the ethics of cloning transplants, arguing that techniques that involved the manipulation or destruction of human embryos were not morally acceptable even when their proposed goal was good in itself (16). This is, of course, a sophisticated ethical argument about means and ends, and an important one. If the means are so unethical in themselves, no goal or purpose can really justify them, and the goal will be corrupted in any event by those means. There will be a different goal or end reached. The question is, is research done on human embryos such a highly unethical action, and the answer is not as apparent as it seemed to Pope John Paul II. Nor is the answer as apparent as it has seemed to government officials who made recent decisions in the U.S. and Britain that allow therapeutic use of human embryos. In the former, the assumption was that the means were highly unethical. In the latter, the assumption was that the ethics of the means didn't matter since the goal/end was highly ethical. The former reflects a conclusion about the moral standing of human embryos that makes them almost equivalent to a developed person. The latter reflects the postmodern rejection of such ethical questions as irrelevant. My argument is that the conclusion concerning moral standing of embryos is not sufficiently warranted, and that the exclusion of this ethical question, reducing ethics to politics or social conventions, is firmly not warranted. In the chapter on abortion issues in medicine, I will develop this more fully. The point I wish to make now is that bioethics has chosen the second approach to all questions about human biotic material because of its political foundation, making ethical analysis actually impossible within its framework. This impacts reproductive medicine because many of its technologies use or could benefit from using research on human embryos and the manipulation of embryos. Many of the technologies use similar techniques to cloning life forms such as Dolly the sheep. The ethical obligation to human embryos is a major issue for human reproductive medicine.

The most sensational aspect of biological cloning techniques is the media case of Dolly. Dolly demonstrated that with enough time, money, and acceptance of reproductive failure, an organism as complex as a sheep could be cloned. What it unleashed was a spate of rash projections and fears, the typical Frankenstein Syndrome which oversimplifies and often misrepresents the actual event. John Dewey made an important point when he said that asking the right questions and identifying the real problems were the primary steps in both science and ethics. In order to do that, an accurate factual description has to be constructed. Yet the history of Dolly was a history of not getting the facts straight. The public was presented with a horror movie scenario by bioethics, something that repeatedly occurs when the issues involve molecular biology and its medical application. What was needed was a reality check.

The researchers who designed the Dolly study asked certain questions and constructed their scientific design to attempt to answer those questions. First, was it feasible to clone more complex animal species? Dolly demonstrated that we had the technology to produce a viable clone, but that it was a very difficult task. The number of failed attempts can be acceptable in other species, but it raises questions about the attempt to clone a human organism, since failure rates would be ethically questionable, perhaps unnecessary human sacrifices. The technique may improve, but for the moment the technology is very inefficient, with many more non-viable or non-functional clones produced than a fully functional one. The failure rate is higher than the natural process of reproduction, although we should remember that the natural process does have a high rate of failure. These failures tend to be aborted spontaneously early in the pregnancy, often too early to know that reproduction has failed. Natural systems are not all the efficient, in fact, are as wasteful and redundant as the environmental interaction will allow to achieve the evolutionary goal of species survival. Cloning techniques waste more, but it is a matter of degree. That raises another important question. Would we have sufficient reason to choose this technology in spite of expense and waste, if not routinely, at least in some unusual circumstances?

The research design was also constructed to look for unwanted consequences, the safety factor that might eliminate harms that needed to be anticipated. Will cloning technology application bring into existence a malfunctioning clone? If this technology were used to solve infertility problems for a couple, this would be an extremely important ethical question as well as a scientific one. The research design was sensitive to such concerns, and Dolly was monitored for any malfunctions or defects. Dolly appeared functionally normal, a typical sheep, and there were plans to ascertain whether she could sexually reproduce normally as well. One concern was that Dolly's combination of new and old cells might result in developmental defects such as premature aging. That caution was repeated many times by researchers, who had a good grasp of what the actual problems of cloning might be. Before any of the science fiction ethics could be done, the question of Dolly's biological normality first had to be asked and answered. The actual problems had to be addressed step by step.

In her third year of monitoring, Dolly's cells are not what was hoped for. In her case, but there are conflicting reports from other cloned organisms, her telomeres are shorter than other sheep of her chronological age. Her telomere length resembles that of her mother and may well be her real biological age. As cells continue to divide during an organism's life, the telomeres at the ends of chromosomes tend to lose a bit of material with each division. This is the normal wear-and-tear of life at any level of organization, but in complex life forms it leads inexorably to aging and natural death. So with her shortening telomeres, Dolly is not only her mother's daughter because she started as a lamb, but she is also her mother's contemporary in terms of cell age, if these telomere trends are accurate. In a sheep with a relatively short life-span and rapid maturity, that may not be as grave as it sounds and Dolly is so far fine. In a human organism, however, it could be a disaster. By the time a cloned human reached the age of reason, that human might also be reaching the age of senility, depending on the chronological/biological age of the cells that were fused together. Given this probable scenario, if we attempted to clone a child for an infertile couple, the potential for serious harm would be very high. More extensive animal research would have to answer the question of the determinant of the biological age of the clone before such technology could be used without accepting too high a risk. In addition, sexual reproduction is an evolutionarily adaptive means of passing on the genetic message, allowing recombination of traits and diversity in the human genome. Moving from sexual reproduction, which is inexpensive, reasonably efficient, a human pleasure, and evolutionarily more desirable, to cloning reproduction would not be a feasible or ethical decision, except in rare circumstances where reproduction could not occur by any other means – even if the telomere question were answered in terms of no premature aging risk.

Granted that all such hurdles were surmounted, there remains the objection that such scientific technologies remove the human from the equation and constitute unnatural interference in a natural order or plan. These objections, while historically centered in religious views of ethics, have transferred through the assumptions of environmentalism into a portion of bioethics thinking as well. From the religious perspective they need to be taken seriously, but I would argue they are mistaken in their description of the definition of the human organism. Humans (whether one believes they are created or evolved) are constructed to be tool makers and tinkerers, and may be well-defined as technologists who attempt to change the status quo for human needs and interests. The technology of cloning is a wonderous extension of human tools for understanding and interacting with natural experience, a very human behavior that need not be devalued. What constitutes an unnatural intervention in this natural experience is also at question. If we look at the systems theory of natural systems, we will not find a steady-state order/balance or plan, but a dynamic process of controlled-state that depends on feedback mechanisms among levels to maintain the functional system (17, 18). The environmentalists and the theologians are both mistaken in seeing a static order or plan, and in viewing the typically-human technological function as unnatural and a threat to that order or plan. Postmodern bioethics has not engaged this debate to any significant extent, however, preferring to place all ethical content and theory aside for the process of coming to a political consensus. For that reason, unfortunately, decisions about cloning have been political, ideological decisions that are made by the representatives of power rather than the searchers for reasoned understanding. The decisions reflect the political biases of the leaders in power, not ethics. Bioethics then assumes an apologetics role in implementing those political decisions through appeals to expert authority and public relations.

Returning to the pope's concern about manipulating embryos for organ supplies in transplant technologies, I will discuss the moral standing of embryos in chapter six. Again, the theological objections must be taken seriously and met with adequate reasoning, rather than putting them aside in favor of a political process ethics. The theologians are correct that they cannot be put aside without also putting aside ethics and classical humanism. There is an important point about respecting our genetic program that should not be avoided in such bioethics discussions, and an appeal to ends justifying the means is the old and dangerous appeal of a totalitarian society. I will disagree with the conclusion that we should not intervene and manipulate human embryos, but the disagreement will address the actual ethical question and not step aside for political process. This question is prior to political theory or ideology, not diminished within the ideology, as postmodernism seeks to do.

The same question actually underlies issues with mapping and sequencing the human genome. How the individual's genome is treated, with respect and with privacy, with personal ownership, monetarily, with consent or coercion, are ethical issues that are slowly being transformed by postmodern bioethics into political issues. Respect is treated as a non-starter, because there is little within postmodern bioethics on which to base a concept such as respect, except the empty phrase that restricts civil rights at the point that other individuals' rights are being infringed. Privacy is defined as a constitutional right, but the legal reasoning behind it is sufficiently ambiguous that it is applied arbitrarily and politically in the legal system (19). Medical records are now routinely collected in centralized government agencies as computerized "data" (having read hundreds of medical charts for various research and legal consults, I can reasonably conclude the validity of that data is suspect) upon which to base insurance and delivery system decisions. The ethical issue of privacy was easily dismissed in the utilitarian name of "better access to health information" and "more efficient delivery of medical care." Postmodern bioethicists supported this very personal invasion of privacy, were in the vanguard in fact of urging implementation of such a system, while some psychiatrists and other medical specialists did try to prevent the construction of such a government bureaucracy of individual medical information (20, 21). In the name of "health," an important need to be left alone, as Arthur Miller sometimes puts it, was seriously compromised. The construction of a total society is incremental.

The individual's ownership of his genetic program and the question of others monetarily benefitting from a person's genes is perceived by bioethics as an important abuse of genetic research and genome projects. However, this is a more compatible issue given the neoleft ideology of bioethics, since it allows for criticism of the profit motive, or capitalism, or corporations, and of victimizing the oppressed. The frame of class warfare and of anti-capitalism grounds ethical objections to patenting genes and genome sequencing. The primary assumption is that such genomic research should be done and controlled by centralized governments, with results shared by those experts within the research and policy elites of governments. The history of the U.S. Human Genome Project is a case example of some of the bad consequences of such an assumption. The government-controlled research was neither the most productive nor the most innovative, and it took the challenge of a researcher who formed his own private research company to unblock the science and create new ideas (22). The results of this project are available to any researcher who wishes to pay for the information, and the techniques developed have sped up the reading of the human genome. The government research provided a forum for bioethics as part of a special component of the Genome Project, incorporating bioethics into the political process and further linking it with a political consensus rather than an ethical analysis.

Ethical issues of consent rapidly become formal, legal consent in current bioethics, and genetic research is no exception. It is, of course, difficult to construct safeguarded consent forms, as Robert Veatch, an expert in bioethics, found during the swine flu incident. Although a bioethicist had helped construct the forms for consenting to the flu vaccine that was rushed into production for fear of an epidemic similar to the 1917 flu pandemic, there were still lawsuits following bad outcomes from the vaccine based on inadequacy of the consent forms. Furthermore, managed care systems require signing away confidentiality to medical information, so economic coercion is also likely. Given the trend to justify coercion in the name of health, a trend supported by current bioethics, it is unlikely that an individual's genetic program will not be used by centralized states to promote health and expand preventive medicine. It is unlikely that personal ownership of one's genetic program, if those in power deem it a necessary resource, will be or can be protected by a politicized bioethics.

Gene therapy which is limited to somatic interventions is not conceptually part of human reproductive medicine, but it is so closely linked with germ-cell interventions, the actual changing of the genome and its implications for future reproduction results, that it is practical to discuss both under this heading. Because medical technology stands a good chance of modifying the genome of egg or sperm, it has been constrained by the same ideological assumptions concerning eugenics and biological determinism that are part of postmodern bioethics. In addition to such political assumptions, there are some valid scientific concerns about risks in changing the genetic program that has slowly evolved with many trial and error experiments in nature.

For example, sick cell anemia, while maladaptive in the homozygous person, can be quite adaptive in the heterozygote, making it difficult for the malarial pathogen to successfully infect and make the person ill. In malarial belts around the world, there is an evolutionary selection for the sickle cell trait (or similar genetic blood diseases) that outbalances the evolutionary deselection. Before human beings have developed a successful vaccine for malaria, it would be premature to use gene therapy to remove the gene for sickle cell anemia. Our knowledge of what is an adaptive gene may in some cases be inadequate to make a determination that holds in the real world. This caveat would not apply to all genetic diseases, but we need to keep it in mind when selecting those for which gene therapy would carry sufficiently low risk. Also, some harmful genes may be linked to very crucial beneficial genes, and removing the harmful gene may upset the linkage and risk the beneficial gene. We are always dealing with systems in biology, and intervening in a dynamic system has to be done carefully to avoid unintended harmful effects. Still, if the genetic disease is severe enough, linkage may not be as great a long-term problem. Nor is it likely that we will successfully perform gene therapy on every human being and therefore completely lose a gene we are selecting out. If a mistake is made, there may be possibilities for recovery. These concerns are ethical cautions, not ethical prohibitions. Finally, some genes that are maladaptive in specific environments may be highly adaptive in a changed environment. Geneticists worry about removing any information from the gene pool, even when it is harmful misinformation. This is a difficult concern to address because one cannot predict future environments and catastrophic changes into infinity, nor predict/guess the gene that could be used to successfully cope with such new environments. It also comes perilously close to an "argument from ignorance" in fallacies of inductive logic. The response has been that we must err on the side of caution in any consideration of changing the genetic program of the human species. If caution means taking time and care in our scientific and ethical actions, that makes bioethical sense. However, if caution means an absolute constraint, resignation and acceptance of a physical evil in the world, or censorship of genetic knowledge, then it does not make bioethical sense (23).

Genetic therapy is currently somatic therapy, and it has only recently been successfully used for various diseases. The media and bioethics have concentrated on the Jesse Gelsinger case, a young man who died during a gene therapy trial, it is presumed because of a fatal immune response to the vector used, an adenovirus. Adenoviruses, which can cause the common cold, are considered safe or mild viruses, but we do need to be careful about vectors, because retroviruses were once considered safe and causing no human disease until Robert Gallo proved otherwise for HTLV-I, and HIV confirmed the proof (24). Finding a safe and efficient vector is an ethical obligation that is not easy to meet, if ethics demands zero risk. Fortunately, ethical theory rarely deals in the search for zero risk, and if zero risk were our goal, a good case could be made that it would be an unethical goal because of the collateral effects of meeting it. Researchers are now looking at adeno-associated virus as a safer vector, but taking risks in some of the genetic disease situations humans confront is often ethical, not unethical.

Gelsinger's death resulted in the expected political bioethics response. The Food and Drug Administration wrote more stringent regulations for gene therapy, and that regulatory response substituted for full ethical inquiry. Following that failure of gene therapy, however, a number of first successes occurred. Laboratory mice with retinitis pigmentosa were successfully treated with gene therapy. Philadelphia Children's Hospital reported that during a safety study for a gene therapy to cure hemophilia, patients receiving a dose so low it had not been effective in experimental animals began producing small quantities of Factor IX which improved their disease and reduced their need for injections. At Necker Children's Hospital in Paris, French researchers treated two babies who had severe combined immune deficiency, and both had normal immune systems after the gene therapy. Gene therapy has also been partially successful in treating advanced skin cancer that previously had no effective treatment. Clinical research, then, is confirming the benefits of somatic gene therapy.

Still, bioethicists have a consensus that advocates that gene therapy be highly regulated and restricted, and that new review committees be formed which would have the power to curtail this research. One of my co-authors, Doris Teichler-Zallen, takes this approach. We disagree on the value of risk-taking and of politically-bureaucratizing ethics. She represents the postmodern ideological consensus in bioethics which prima facie assumes that genetic intervention will be unethical unless strictly-controlled by government institutions, and that political forms of analysis and activity are the tools of bioethics.

Any risk in ideologically-suspect fields becomes unacceptable risk for political bioethics. However, I wish to argue that an adequate understanding of science and human history makes clear enough that risk taking is required for growth and optimal functioning. A learning program requires mistakes, human immunity requires exposure at an early age to pathogens, human evolution requires mutations and variances, home territory requires physical defense, and wisdom requires failures and fools. Free functioning implies exposure to the dangerous and unknown, and medical progress results in deaths and complications. We can try to limit the severity or amount of the risk within reason, but too much restriction brings things to a static decline. All medical knowledge through history, as far as I can see, required some risk taking. In genetics, risk taking has not created a major catastrophe in spite of Doomsday Scenarios. If we strip away the fantasies of eugenics and the science-fiction, we see failures and a small number of deaths. If we choose to institute political regulatory systems to slow down genetic discoveries and applications, we need to look at the reverse side of the coin. Many people will then be deprived of applications that could have relieved their morbidity or prevented their mortality. A balance is the ethically sound approach, but not a utilitarian balance. We will need to look at systems theory methodology in Chapter 5

Let me give a hard-headed example. Suppose it were true that polio vaccine experiments in Africa had spread HIV from monkey to man. If that were the case, should researchers have refrained from testing the polio vaccine, and waited until they knew all the viruses that infect monkeys, the statistics on risk of species jump to man, the effects these viruses would have on humans, produced a cure for every possible infection that could develop, or wait until a new medium for vaccine were produced? Or, should researchers have taken the risk and done field trials? I would ethically opt for testing the vaccine, because the risk of polio and the risks of HIV are similar enough. A known evil such as polio, I would argue, has to be fought even when the risk is not zero. Given the nature of the biosphere and how natural systems work, it is also very likely that HIV would have found other ways to make the jump to humans (as it did). I would ethically conclude that refusing to take the risk is less ethical than taking the risk.

To the extent that legislation and regulation pursue zero risk and censor research, the unacceptable status quo is reinforced. Laws and regulations may be ideologically appealing, but they generally fail at their purposes and restrict advances in knowing the world around us. Instead of a postmodern bureaucratic ethics, we require a theory that understands the nature of risk taking and its relationship to growth and adaptation, to learning and exploration. Geneticists like to quote Sophocles that "it is but sorrow to be wise when wisdom profits not," but that misses the reality that wisdom almost always profits and that forbidding understanding and application almost always harms more than benefits. We have placed our evolutionary bet as a species on a learning program. It may in the end be a losing bet given the realities of species extinctions, but we are doing well so far at genetic roulette. Learning and risking are in our genes.

REFERENCES

1. Wilson, E. O. (1992). The Diversity of Life. Harvard University Press, Cambridge, MA.
2. Ridley, M. (1999). *Genome: The Autobiography of a Species in 23 Chapters*. New York: Harper Collins.
3. Schreuder, C. (11/23/1997). *Septuplets' birth resurrects debate on fertility treatment*, pp. 1, 24. Democrat and Chronicle, NY: Rochester.
4. Stenberg v. Carhart. (6/28/2000). *Excerpts from Court's Decision on Nebraska's Abortion Law*. New York Times.com.
5. Clements, C. D.(1982). *Medical Genetics Casebook: A Clinical Introduction to Medical Ethics Systems Theory*. Clifton, NJ: Humana Press.
6. Schreuder, C. (11/23/1997). *Septuplets' birth resurrects debate on fertility treatment*, p. 24. Democrat and Chronicle, NY: Rochester.
7. International Humanist and Ethical Union (2000). *Humanism for the World*. IHEU, London.
8. Spengler, O. (1928). *The Decline of the West:Perspectives of World History*. New York: Alfred A. Knopf.
9. Arnett, E. C. (12/17/1998). *Limits on multiple births urged*. Democrat and Chronicle, NY: Rochester.
10. Elster, N. R. (1999). Legal and ethical issues raised by assisted reproudction. In: R. B. Edwards & E. E. Bittar (Eds), *Advances in Bioethics* (pp. 269–284). Stamford, CT: JAI Press.
11. Clements, C. D.(1999). Old women, pregnancy and postmodernism. *Free Inquiry, 19*(4), 35–36.
12. Calabresi, G., & Bobbit, P. (1978). *Tragic Choices: The Conflicts Society Confronts in the Allocation of Tragically Scarce Resources*. New York: Norton.

13. Meadows, D. H., Meadows, D. L., Randers, J., & Behrens, W. W. (1972). *The Limits to Growth*. New York: American Library.
14. Winner, L. (1986). The political philosophy of alternative technology. In: A. H. Teich (Ed.), *Technology and the Future*. St. Martin's Press.
15. Mayr, E. (1982). *The Growth of Biological Thought: Diversity, Evolution, and Inheritance*. Cambridge, MA: Harvard University Press.
16. Los Angeles Times. (8/30/2000). *Pope condemns cloning*. Rochester, NY: Democrat and Chronicle.
17. Clements, C. D.(1995). Stasis: The unnatural value. In: R. Elliot, (Ed.), *Environmental Ethics, Oxford Readings in Philosophy* (pp. 215–225). New York: Oxford University Press.
18. Clements, C. D.(1982). *Medical Genetics Casebook: A Clinical Introduction to Medical Ethics Systems Theory*. Clifton, NJ: Humana Press.
19. Griswold v. Connecticut. 381 U.S. 479 (1965).
20. Gostin, L. O., Turek, B. J., Powers, M., Kozloff, R., Faden, R., & Steinauer, D. D. (1993). Privacy and security of personal information in a new health care system. *JAMA*, *270*(20), 2487–2493.
21. Sweeney, L. (1997). Weaving technology and policy together to maintain confidentiality. *Journal of Law, Medicine & Ethics*, *25*, 98–110.
22. Wright, R. M. Balint, J. A., Porter, I. H., & Shelton, W. N. (1999). Ethical issues in genetic research, testing, counseling, and therapy. In: R. B. Edwards & E. E. Bittar (Eds), *Advances in Bioethics* (pp. 171–213). Stamford, CT: JAI Press.
23. Hubbard, R. (1982). Some practical and ethical constraints on genetic decisions about childbearing. In: D. Teichler-Zallen & Clements, C. D. (Eds), *Science and Morality: New Directions in Bioethics* (pp. 37–47). Lexington, MA: Lexington/Heath.
24. Landau-Stanton, J., & Clements, C. D. (1993). *AIDS, Health and Mental Health: A Primary Sourcebook*. New York: Brunner/Mazel.

5. INFECTIOUS DISEASES: ETHICS, EXPERTS AND POLICY IN AIDS, vCJD, EBOLA VIRUS, WEST NILE VIRUS

THE POTENTIAL RISKS OF A PUBLIC POLICY MISTAKE

In the early 1980s, I began to hear the first murmurings at the medical school of a new infectious disease that was causing a rare form of cancer among homsexual and was severely compromising their immune systems. Some medical professionals were calling it "GRID" for gay-related immunodeficiency. Later, when I became coordinator for training in our AIDS Training Project (NIMH), research indicated that there had been mysterious deaths reported out of Africa, one a Danish woman who was a surgeon at a missionary hospital, another an airline worker. In North America, also as far back as the 1950s to 1970s, a similar cluster of symptoms new to medicine was being reported in non-homosexuals who had returned from Haiti. One, a French geologist, had received blood transfusions and died of cryptosporidiosis. The other, a merchant marine from the U.S., died of pneumocystis carinii pneumonia in 1959. These very small blips on the radar screen quickly disappeared into the collections of rare cases of unknown etiology, to silently wait for ten or twenty years before their significance was understood (1, 2).

A few infectious disease specialists were arguing that this new immune deficiency disease seemed likely to follow a model of transmission similar to hepatitis B, and the Centers for Disease Control had assigned Sandra Ford to sound the alert if request for pentamidine to treat pneumocystis carinii pneumonia in a hemophiliac came across her desk. One did. Hemophiliacs, who required blood clotting factor from a blood product made up of many donors, were chosen by the CDC for surveillance because they were similar to the canary in the coal mine (3, 4).

By 1982, the CDC had tried to warn about transfusion as a means of transmission of this fatal, non-treatable disease, but in January of 1983, the heads of blood banks still opposed screening donors and the Food and Drug Administration had not acted decisively. It wasn't until March 3–4 that pressure resulted in the U.S. Public Health Service requesting that risk groups refrain from donating blood and that guidelines for screening donors were put in place.

Still, on March 7, 1983, at a New York University AIDS conference, experts were still claiming that the transfusion risk evidence was lacking and only inferential – in other words, a case of transfusion-transmitted AIDS/HIV had not been "documented." The term "documented" is another code word that means we have not absolutely proved the transmission route by having a specific case where step-by-step the transmission was identified beyond any doubt. Experts at conferences still maintained a premature consensus about transmission. In the homosexual community, where political and social pressure was being used to resist screening and testing (when available), and closing of bath houses where multiple sex partners could easily spread the disease, Larry Kramer, a gay activist, was a voice crying in the wilderness. The Lambda Legal Defense Fund threatened to legally block HIV testing, the New York City Health Commissioner ordered that no lab in the city do HIV tests except for research purposes, and physicians were called fascists for recommending voluntary HIV testing (5, 6).

In actuality, by 1982, the handwriting was on the wall about AIDS and risk groups, if a policy maker wished to read it. AIDS was rampant among New Jersey drug addicts, and pediatric AIDS cases (though not diagnosed as such) were in the dozens in a hospital in New Jersey. An elderly hemophiliac had died of pneumocystis carinii pneumonia, and a young baby transfused because of Rh factor had immune deficiency develop after the transfusion. It was found that the blood donor had lymphadenopathy and an eye problem.

None of this information seemed to matter for public policy. Activists, professionals, and public health officers were asking for impossible absolute documentation (the sort demanded of the documentation of the chain of hand-over of DNA evidence in the O. J. Simpson case) before they would act on the high probability of HIV being transmitted through blood. Policy seemed to be requiring absolute proof rather than sufficient scientific probability. Policy experts made a choice on policy that over-emphasized proof of transmission and under-emphasized erring on the side of safety. Opposing views were suppressed at conferences and meetings in either subtle or obvious ways, and the premature consensus remained in place. It was not a good day for the scientific method. (7, 8).

The results of this hardened consensus still haunt us. The French Association of Hemophiliacs estimated that in 1991 1,200 of 3,500 hemophiliacs were positive for HIV. In the U.S., 12,000 individuals became infected with HIV before testing was instituted or screening begun, and 70% of all U.S. hemophiliacs during that period became infected with HIV. Results across the globe reflected the human cost of demanding absolute proof (documented transmission) of transmission before initiating screening of donors. (9, 10).

There has not been a strong-enough case made for the ethics of public policy, because the same pattern was in danger of re-emerging in 1999 in the arguments over screening blood donors to prevent what has been called variant Creutzfeldt-Jakob (vCJD) disease. I had suggested in my ethics columns for *The Medical Post* that this family of progressive neurological diseases be called spongiform encephalopathies (SEs) to indicate that the pathogen, probably a misfolded protein, has often crossed species through a variety of transmissions and manifested itself in modified clinical signs from species to species. The literature is now beginning to refer to this infectious disease entity as transmissable spongiform encephalopathy (TSE) which can help clarify policy and public understanding.

I have always been intrigued with TSEs or CJD. When we conducted our AIDS Training Project, we discussed CJD and its iatrogenic transmission through organ transplants, human growth hormone injections, surgical instruments, and most likely, blood. As a result, I followed the sporadic outbreak of vCJD in Great Britain very closely, picking up the first professional reports in 1986. I also watched the repeating public policy mistakes that had been made with AIDS/HIV being made with vCJD. In the beginning of the outbreak and the subsequent epidemic in the British cattle herds, officials inaccurately reassured the public, the media characterized BSE as specific only to cattle, blue-ribbon committee reports were kept secret and only became public when a member's list of human cases was faxed to the wrong fax number, blood and other organ banks resisted any possibility that it could be transmitted through blood or other donations, and experts kept talking about the lack of proof (no documented case) of cattle to human transmission. This was done despite veterinary knowledge that the SEs had in the past spread from sheep to mink and had likely spread currently from sheep to cattle. It was done despite Dr. Carleton Gajducek's award-winning work on the spread of another SE, kuru, from human to human most probably through ritual funeral cannibalism; and the pediatric evidence that human growth hormone, infected with CJD, had infected some children treated with it. In addition, government health officials and agricultural officials fought to save the British beef industry with the same fervour that San Francisco health and public officials had fought to save the gay bath houses in the early days of AIDS/HIV.

Despite the repeating pattern of resistance by some experts, however, North American health officials may have learned from the AIDS/HIV story. Unlike that policy history, a reasonable screening program for blood donors has been put into effect fairly early in the history of this new outbreak, screening out potential blood donors who spent six months or more in the U.K. between 1980 and the present. Since there is still no reliable blood test for vCJD or TSE, as was the case shortly after donor screening for HIV, screening donors remains the only public health measure available in 2000.

What complicates the response to TSE are the political and economic factors in public policy decisions, and the relatively easy way that policy-making can become so politicized that even the science becomes affected. I want to take a look at some specific ethical issues in this recent infectious disease problem.

One key ethical question is *conflict of interest*, something the expert can become aware of and make known to the public, but is generally not possible to eliminate in the real world in which ethics needs to be done. In this case, the response of experts with a potential conflict of interest because of involvement with blood banks has mimicked the response of 18 years ago. Some still question the transmission of vCJD/TSE from eating beef, although those questions should long ago have been laid to rest or, given the history of scrapie (the TSE in sheep), never asked in the first place. To suggest there is no evidence of transmission between cattle and humans is to deliberately overlook the long history of the TSEs and their inter-species transmission through injestion. There has always been more than enough scientific evidence that TSEs are inter-species transmissable no matter what the species, and there is currently solid scientific evidence of transmission to other species in the British experience (to humans, cats, zoo animals, e.g.). As was the case for human growth hormone contaminated with the CJD agent, it takes time for monkeys or chimps to develop the disease after injection, since this is a "slow" agent. Gajdusek's chimps finally did develop a form of SE following injection with CJD-contaminated material, but public health agencies did not wait for that unnecessary absolute proof before pulling all three batches of human growth hormone from the world market. In doing so, those agencies saved lives that would have unnecessarily been lost. Furthermore, if TSEs can be transmitted by organ donation, growth hormone injection, and cuts in the gloves of neurosurgeons during brain operations, it is reasonably clear that blood transmission becomes a high probability. We should not have to answer the silly objection of "no proof" at this stage. But we have to, because experts with obligations to their particular systems, for example, blood banks, ministries, global free trade organizations, etc. have other values that represent a conflict of interest. That policy on the blood supply was recently scientifically confirmed in *The Lancet* and reported in the New York Times (11), and has resulted in Canada now screening out blood donors from both the United Kingdom and France who have spent a cumulative six months in one of these countries since 1980.

In July of 2000, the U.S. government planned to seize and destroy 400 sheep in Vermont over the objections of the owners, because some imported from Belgium had tested positive for scrapie, the sheep TSE that infected British cattle and led to vCJD (12). Yet the New York Times reporting of the story shows how easily deception can occur. The news story asserted that BSE (beef SE) had not been known to be passed from cattle to sheep except in the laboratory, but this only means there is no specific documented case that was looked for in the natural setting. We know that TSE crosses almost all species. We have known that for some time, with a high degree of scientific probability. The report says agricultural experts choose to err far on the side of caution, but it is not erring far on any side to recognize the extensive body of evidence about TSEs. The spokeman for The Department of Agriculture characterized this as acting with an "abundance of caution," but that too is misleading. It is normal, sensible caution for public health, not an abundance of anything. This represents another tension in the ethics of public policy, what I have called in articles the "Don't Panic Scenario" vs. the "Doomsday Scenario." The experts here are influencing the reporting of the story closer to the Don't Panic Scenario. The large ethical question for experts and public policy is *how and where on this continuum of Don't Panic Scenario vs. Doomsday Scenario one places the problem.*

In this atmosphere, it is easy for the owners to call the seizure "Gestapo tactics" that are based on "questionable" science, when in fact they are not. One owner wanted seizures delayed until it was determined whether the sheep really had TSE, not understanding that scrapie *is* a TSE. The news then goes on to list 180,000 British cattle that were infected with BSE through feed that included the rendered brains of "diseased" animals, not bothering to mention that the disease was scrapie. Even a senior staff veterinarian at the Department of Agriculture claimed the sheep had scrapie, but that scrapie had appeared in sheep for centuries and not affected humans. This is also misrepresentation. With the exception of the British, countries have tried to eliminate scrapie from their sheep flocks with aggressive means, and there is some evidence that ethnic diets using sheep eyes can transmit sheep SE to humans in the form of CJD. The logical inference, based on the ease with which SEs cross species, is very strong, and the scientific evidence of scrapie infecting other species such as mink, cattle, and zoo animals is also strong. There is no good reason to believe humans would be an exception, unless the experts have chosen an extreme Don't Panic Scenario. For one of the owners of the sheep to say that there "has never been a sheep in the whole world that has gotten mad cow disease," is to show how misinformed the public has been by expert policy choices on the continuum of Don't Panic Scenario to Doomsday Scenario. Sheep have been infected in the laboratory with BSE and carry their own TSE (scrapie) as well.

Other evidence is available to make an expert choice of the Don't Panic Scenario unethical. In French zoos, all 18 apparently healthy lemurs were infected with TSE/BSE, probably as a result of feed manufacturers adding British beef to their products until 1996, an unconscienably late date. In France, BSE is spreading, requiring destruction of some cattle herds. It has been discovered in native-born cattle in 10 European countries (e.g. Switzerland, Belgium, the Netherlands, Portugal, Germany, and Denmark). Moving on the continuum away from the Don't Panic Scenario and halfway to the Doomsday Scenario, France is beginning random testing of all cattle who died showing suspicious behavior. The European Union is urging broader preventive measures throughout its 15 member countries. Some experts are beginning to consider that in addition to injestion of contaminated food and maternal-fetal transmission, there may be an unknown third form of transmission. France is refusing to accept shipments of British beef, which will result in a legal battle with the European Union that insists British beef is as safe as any, so both sides are on opposing points on the continuum, with the EU far over to the Don't Panic point. The U.S. agrees with France on the risk, also banning importation of British beef. The U.S. itself has problems with TSEs in its wildlife in several western states (calling it chronic wasting disease). In one county in Colorado, 15% of deer are infected, and elk throughout western states have had TSE, as reported by Gajdusek as long ago as the 1980s (13). In August of 2000, in Spain, authorities destroyed 2,400 sheep after detecting Spain's first cases of scapie, and will destroy another 400. In the ongoing public relations confusion, this news story said scientists now knew BSE (popularly known as Mad Cow Disease) did jump species barriers – something that actually should have been reasonably inferred from data fifteen years ago.

This confusion represents the ethical tension experts have between *informing and involving the public*, and *causing more harm by premature or unjustified precautionary actions*. This ethical analysis and choice is difficult enough to make if the expert and the policy institution are *operating on standards of objective, rational thinking, as corrected for bias and conflict of interest as possible, as open to new experiences as possible, as honest with public information rather that public-relations deception as possible*. When, as is more and more the case, the policy institutions are politicized and responsive to ideological agendas and power strategies (as I will show is the case in AIDS/HIV, the TSEs, Ebola virus, and West Nile virus), public policy in infectious diseases may ethically fail the human population.

Continuing with the TSE and AIDS/HIV situations, some experts working on TSE policy question the estimates and projections, taking us into the arena of *risk assessment and risk/benefit analysis*, another ethical question in public health policy and policy in general. It is understandable that a projection of vCJD that ranges from a few dozen to several thousand or hundred thousand is questionable. Reported deaths from vCJD range: 59; about 50; more than 35; 56 in Britain, 2 in France, 1 in Ireland; to 82 in Britain and counting (14). In the beginning, the special British committee knew of human deaths numbering in the teens, but kept this information from the public, thus stepping into *the ethical question of involving the public in policy or deliberately and subtly deceiving the public and keeping information restricted.* But a mistake by one committee member that sent the list of human cases to a private home fax and from there communicated to the press finally involved the British public. Now, reports are made about trends, and a recent one is disturbing. In the first half of 2000, 14 Britons died of vCJD, as many as died all the previous year, and 5 others were known to be dying (15). One estimate, if the trend continues and we are seeing an epidemic in its early stages, is that 500,000 Britons could die over the next 30 years. Whether the numbers hold steady or fall, some experts now project hundreds or thousands of people dying of vCJD. The acting director of the Spongiform Encephalopathy Advisory Committee which advises the British government, Dr. Peter Smith, is not ready to make a projection, feeling it is difficult to know what the new numbers mean at this point. He then gave the number of deaths as 74 in Britain, 2 in France, and 1 in Ireland. He found the increase in deaths in 2000 surprising, but not necessarily portending a large epidemic (16). A probable cluster was also identified in the small village of Queniborough north of London. Four young adults had died from vCJD in 2000, and a fifth was near death. Epidemiologists were investigating 10 local slaughterhouses where specialty meats such as sausage could have been eaten by children. Another projection from some British scientists is that 75 people will die of vCJD in the next 20 years, while others project 80,000 (17).

From these figures which are all over the board, it is easy to see that risk assessment and risk/benefit analysis will be difficult for policy planners at best. The exposure of the British public was high, but that is only one factor in risk assessment. The infectivity of vCJD or other TSEs is not sufficiently known.

The length of time before symptoms, sometimes over 20 years and often mistaken for other neurological diseases such as Alzheimer's, gives the pathogen an advantage. By the time symptoms appear, it is too late to stop the exposure and prevent an epidemic. Tests are being developed that could alleviate this problem for policy makers, and they are badly needed. But in the meantime, policy must be made with an *uncertainty factor* that is uncomfortable. If history is considered, however, there is an example that could help make risk assessment and evaluation. Experts may not know the extent of potential transmission at the moment, but they can know the risk. Experts can use the scientific knowledge of what happened to the Fore people of New Guinea when they engaged in ritualistic funeral cannibalism. The usually sporadic nature of another variant CJD, kuru, suddenly became an outbreak that swept through the people most engaged in the ritual, women and children. The women prepared the bodies, covering their hands with blood, brain tissue, and other body tissue. The children ate with the women. The men, who didn't do the preparation and ate little of the body were less affected. Instead of the normal 1–2 per million population for TSE, the women and children of the Fore were decimated. Was this a false correlation between injestion of the dead and kuru? When the cannibalism practice was stopped, the rates of kuru dropped rapidly, reaching the sporadic level quickly. Was there no evidence for SE transmissibility from body tissue? There was good field evidence, though not laboratory documented transmission, but that field evidence was sufficient, given the severity of the risk consequences, to make certain policy preferences. Should the Fore have been allowed to donate blood? Given strong indication of other organ transmission, the risk of blood donation, while not documented, needs to be seen as high-risk. Blood transmission, in addition, can be one of the swiftest means of transmission in a population, and we should remember how quickly HIV spread among the hemophiliac population. It would be unethical risk-taking to have used Fore blood donors, given the amount of information available to experts and the severity of the consequences. This would not have constituted European "discrimination" against the Fore, anymore than screening out homosexuals or Haitians would have constituted "discrimination" against them. That is a meaningless term in these policy choices. Would the risk perception be skewed because there was no documented blood transmission of kuru in the Fore? Policy would not have to wait for such documentation to be obtained, and I would argue it would have been unethical to wait for such a demand of documentation. This chapter will discuss the components of "risk" in more detail, but for now it will argue that "risk" is made up of three important factors: (1) the statistical, probability figure of risk, (2) the severity of the consequences of taking the risk, and (3) the individual's control of choice to take the risk or the imposition of the risk without individual control (18, 19).

Unfortunately, currently with vCJD we don't quite know the probability number and the projections are not scientifically established. The severity of the risk is the highest possible in many senses: an invariably fatal, untreatable disease that progessively takes away that important human function of thinking, perceiving, and interacting with others. Without screening of blood donors as a first step, the individual has no control over whether to take the risk or not, short of refusing a blood transfusion even in life-saving situations. All this adds up, using the equation of risk, to a very high risk perception for any rational being. In terms of public health, if it could be prevented, no individual should have to assume such a risk.

Other experts, making an evaluation of cost/benefit rather than risk/benefit in the utilitarian ethics sense, want to argue that we should focus on more common blood pathogens before we invest the staff, time, and money resources on TSE/vCJD. This is a divide-and-conquer approach that has no logical merit. We should be doing all of the above, not economically pitting TSE against other pathogens. We should be working on developing artificial blood, because human biotic material is always a presumed biohazard. Also, as I have argued, the potential risk consequences of TSE/vCJD are much worse than most other pathogens, many of which we have at least a chance of treating. TSEs destroy the brain.

Only time will fill in the probability numbers and allow us to project the real consequences of the actions of a cattle industry that foolishly fed cattle the remains of sheep who had died of scrapie. That blunder is almost inconceivable, given what we should have known about TSEs. But before the cattle industry is singled out too harshly, public policy experts should not commit the same error by accepting blood donations from people who were exposed to British beef or beef by-products as safe donations. From a full and scientific cost/benefit analysis, the 3% loss in blood donations is a small price to pay. Even if it turns out that very few people actually die of TSE/vCJD and the outbreak remains sporadic, it is still a small price to pay for infection control. Ethically, the public should not have to take such a risk with blood transfusion. An unnecessary utilitarian sacrifice of individuals is unethical, even when the sacrifice is only probable and not absolutely demonstrated to occur. Few things in life are ever absolutely demonstrated.

The ethics of public policy needs to determine how much demonstrability makes sense in a given situation, and to make that determination fully aware of conflict of interest and the requirement of *balance* in a system. When policy becomes politicized, when public health becomes politicized, unnecessary harm occurs.

AIDS/HIV AS A POLITICIZED DISEASE IN POLICY AND BIOETHICS

Hiram Caton raised the question of policy making and funding for AIDS/HIV (20), and this history of this disease both in the U.S. and globally makes little superficial sense unless we look at how AIDS/HIV has always been a politicized disease. From the deception that AIDS/HIV was a serious problem for the general population in the U.S. in order to attain more funding for research and treatment (21), to the decision not to use adequate public health measures in Africa that allowed this disease to extend to new areas of the continent and grow within the first-affected areas, health policy has been heavily influenced by political ideology and power.

AIDS/HIV first came to medical attention within the homosexual community and coincided with a Gay Rights movement that stressed political power and sexual freedom. Although Randy Shilts documented that homosexual physicians in San Francisco in 1980 had warned that the developing lifestyle could become a highly successful vector for known infectious diseases and a potential disaster for any new lethal disease, he was on the losing side of a political split that developed in the homosexual community (22). Any criticism of the multiple-partner lifestyle was seen as traitorous to gay civil rights, as fascist, homophobic (23). When the new immune deficiency disease surfaced, homosexual legal and political organizations blocked public health measures that would have protected their own community, such as closing bath houses, testing, and avoiding anonymous multiple sex partners. Certain politically-uncomfortable studies were suppressed, and charges of public health discrimination against homosexuals made these measures difficult to implement. A presidential campaign even suggested that if one only voted in an ideologically-proper way, this disease would be resolved. When the danger to the homosexual community could no longer be denied, because biology can only be denied for so long politically despite postmodern denial of accurate reality, this community embraced AIDS/HIV as the opportunity it was to make homosexuals full members in the human community.

With the disease so closely related to the homosexual community, other risks groups were missed, such as women, children, and drug addicts. In addition, there was another political problem involving not only discrimination but racism, as the underclasses in the cores of central cities became involved in the disease. The same political difficulty, exemplified in both the racism and colonialism assumptions the neoleft made in public health and bioethics, restricted policy makers in both the U.S. and globally in the UN.

The same problem of projection of HIV infection that had occurred with projection of TSE infection in Britian also occurred in Africa. There are problems in the tropical areas with using ELISA testing (false positives), with the cost of polymerase chain reaction testing that is more accurate but medically-intensive and expensive, and with using skewed populations such as the sex trades. The rates of infection have been consistently estimated as quite high (24). Because the U.S. had made the political choice of preventive education rather than contact-reporting and other traditional public health measures, that choice was also reflected in the UN and the global response to AIDS/HIV. This disease, in the past very rare in Ethiopia, is now spreading rapidly in that country, but is mainly being managed with a policy of preventive education because the more traditional public health measures violate the postmodern bioethics consensus. There are reasons why preventive education would never be adequate, as it was inadequate for syphilis (25). The current pause in AIDS/HIV disease in the U.S. homosexual community may be due more to the development of the treatment cocktail than to radical changes in lifestyle, and if resistance develops to these antiretrovirals, the health picture will again grow dim. In Africa, the situation is even more complicated. Many Africans, including professionals, believe that HIV is transmitted through insect vector as well as other means, and are fatalistic as a result. Also, other serious diseases such as malaria or tuberculosis will likely prove fatal to many Africans before HIV infection becomes a serious concern. Again, there is a sense of fatality which can easily submerge preventive education. Grmek has listed migration, civil war, economic dislocation, and the breakdown of tribal or village constraints on sexual behavior as major factors in the history of AIDS (26). These are major forces that are not so ammenable to ideological assumptions about prevention and education as the bioethics consensus would like to believe.

We are fortunate that AIDS/HIV is not a pandemic equal to bubonic/pneumonic plague, treatment-resistant tuberculosis, or the 1918 influenza pandemic. Otherwise, the political consensus of bioethics would have extremely harmful effects. Its political emphasis on civil rights and free choice makes it difficult to use standard public health measures such as identifying carriers, quickly isolating an outbreak area, notifying contacts, and restricting certain behaviors. The too-hasty conclusions about transmission and the unwarranted assurances of safety make it difficult to control pathogens. Attacking risk assessments or infection control requirements as hysterical, discriminatory, racist, or colonialist-imposition of Western science censors information and research. Our research group was told at the beginning that we should take care using the expression "risk groups," since it was deemed politically insensitive by funding agencies. In keeping with ongoing research, we wished to use the term "core populations" as we were further along in the work, but this time we were advised that this term was now deemed politically insenstive by the agencies and we should use "risk groups" instead. "Core populations" is actually an important concept in research as it applies to U.S. cities, and we had data based on zip codes that distinctly showed the city cores. It was just that we were restricted from using it based on ideology.

This brief history of work with AIDS/HIV illustrates a number of the ethical concepts in public policy and expert obligation. Deception of the public did occur, false assurances did occur, risk assessment was not scientific but political, balance was not a goal, conflict of interest was common, and finding the accurate point on the continuum between Don't Panic and Doomsday has yet to be achieved. Under such circumstances, doing an ethical and rational risk/benefit analysis is extremely difficult.

DOOMSDAY AND EBOLA VIRUS

In Chapter 1, I discussed Ebola virus under the section of identifying the pattern of political activism in current bioethics. Ebola virus is a filovirus that has never started a full-scale epidemic even in fertile conditions for such an epidemic. It has been confined to sporadic, iatrogenic and sometimes natural outbreaks that appear to be self-limiting. It is one of the viruses causing a hemorrhagic fever, many of which are found in tropical areas around the world. The fatality rate is variable, from 53–90% fatal in the 1976 outbreak in the Nzara cotton factory and Yambuku Hospital in what was then known as Zaire. A form of this virus was identified in 1967 in a laboratory in Marburg, Germany, when lab workers became infected from imported Green Monkeys (Vervets) from Africa, and over a quarter of them died. Still, very few people had heard of Ebola virus even during the Nzara outbreak in 1976 and a repeat of that outbreak in the same factory in 1979 that killed 65% of those infected (27).

In 1989, a research laboratory in Reston, Virginia, a suburb of Washington, DC, received shipments of monkeys from Asia and these monkeys began dying. The expert researchers at the U.S. Army research laboratory for infectious diseases at first treated the monkey deaths as a simian hemorrhagic fever and were casual about the way materials were handled, although Marburg should have given them pause. The pathogen was finally identified as Ebola virus and at that point extreme precautions were taken. Ebola requires a biohazard level

4 containment, the highest. Three workers were infected, but fortunately none developed major symptoms, and this strain did not appear to be as virulent as Ebola Zaire. As the army began killing monkeys and sterilizing the premises, no one in Reston was informed. This operation was kept from the public until it was over. A reporter was then given access and did first a magazine story, and then, a book on Ebola virus (28).

The academic book that resulted from the Rockefeller University and National Institutes of Health conference presented a careful and nonpolitical description of emerging, resurging, and resistant pathogens, including Ebola virus. The public, however, is more likely to recall and have read Richard Preston's *The Hot Zone*, or watched the *48 Hours* presentation, *The Danger Zone,* or followed the news reports of another outbreak of Ebola virus in Kikwit, Zaire in 1995. Toward the end of that outbreak, about 315 people had become infected and 77% died. Still, even with the excellent conditions Kikwit presented for an epidemic, this outbreak did not spread to Kinshasa as feared, and even in the presence of a civil war later in Zaire which changed its name to the Congo, the Ebola virus has yet to become a spreading epidemic. However, if the public were asked about the danger of Ebola virus, most would have the perception that it could be on any plane headed for the U.S. or other countries and could cause a lethal, nationwide epidemic or even a doomsday world pandemic. To add to this strange public perception of Ebola virus, Preston had suggested it was caused by humans intervening in the rain forests and, in a sense, was the revenge of the rain forest. The public also thinks, as a result, that human beings are the real cause of Ebola fever for having committed a sin against environmentalism. The devaluation of the human species; the faith in a perfectly-balanced, steady-state system of the biosphere; and the rejection of the nature of humanity as a learning program developing tools/technology; combined with an antagonism toward evolutionary explanations and genetic determinism – all constitute the Green politics and pseudoscientific environmentalism of this postmodern ideology. To this almost religious character of current bioethics has been added the secular equivalent of apocalyptic thinking, as McGinn briefly notes (29).

Again, rational public health policy is hard to achieve under these limitations. I would not argue that the increase in funding for surveillance of infectious diseases was a bad policy conclusion, because I think in fact such funding is justified, but rather that the discussions leading to the conclusion were bad discussions. Ebola virus was not correctly assessed in the context of other risks or its own history, and the public was manipulated by purple prose and moonsuit images. The responsibility for releasing such pathogens was inaccurately ascribed to human interventions alone, rather than seen in terms of microbial history over long periods of time, and the solution was presented as less human activity when that is not reasonably possible. The most disturbing consequence of all this is the targeting of individual human beings and their personal activities as the best public health response to infectious diseases, the only response that public health policy experts activate in many situations. In essence, the public is told it is their doing, and they are on their own.

THE POLICY RESPONSE TO WEST NILE VIRUS

One of the best recent case examples of converting public health policy to public relations and individual burden is West Nile Virus. This virus is a flavivirus related to some rather nasty viruses found around the world: Japanese encephalitis, Murray Valley encephalitis, St. Louis encephalitis, Stratford virus, and Yaounde virus. West Nile virus was unknown in North America, however, until 1999. It was first identified in the West Nile District of Uganda in 1937, so it is not one of the newly-discovered viruses like Junin or Ebola. There was an outbreak in Israel in 1957 among elderly patients who suffered severe human meningoencephalitis (inflammation of the brain and spinal cord). Recent outbreaks of WNV were in Algeria in 1994, Romania in 1996, The Czech Republic in 1997, the Congo in 1998, and Russia in 1999, so it is a spreading virus, mosquito-vectored (and possibly tick-vectored) and carried by infected birds. In the 1999 outbreak in North America, humans and horses were infected and developed encephalitis. From August 22 to October 17, 1999, 25 horses in New York had neurologic signs indicating WNV infection (ataxia, head tilt, muscle tremors, inability to rise, e.g.), and nine of the horses died.

The 62 human WNV infections came from either New York City or Nassau or Westchester counties, and seven died. A serosurvey of the New York City area affected found 19 of 677 persons tested were positive for WNV.

Small animals can also become infected. A kitten in New Jersey was documented to have WNV, and a serosurvey of healthy dogs in New York City and Nassau County showed a 5–11% infection rate. Of the small number of dead birds actually tested (about 758 of 17000), 32% of New York birds had WNV and 77% of Connecticut birds, with the overall percent positive 39% (30, 31)

Spraying with malathion was done in New York in 1999, but criticized by environmental activists. The spraying, unfortunately, was done very late in the season and missed the chance to contain the outbreak. From all indications, this was a small outbreak, with the possibility that aggressive public health measures of containment might have prevented the virus from getting the critical mass necessary to become endemic in North America. Because the public was reassured that WNV was not transmitted from person-to-person or directly from birds and pets, and were told that the clinical symptoms were generally not life-threatening (3–15%), and that their likelihood of becoming infected from a mosquito bite (one) was about 1 in 1000, there was little support for widespread spraying. Although the CDC recommendation was that following active sampling and detection of WNV, a rapid response for mosquito control should be implemented (application of adulticides and larvicides), the response was neither rapid nor aggressive. The risk assessment in the public mind was that spraying dangerous chemicals was higher risk than infection and encephalitis. The reassuring words by now are familiar: "There is no documented evidence of person-to-person, animal-to-animal, or animal-to-person transmission of West Nile virus" (32).

The opportunity to contain the outbreak probably vanished over the winter, as WNV was contained in adult Culex mosquitoes that routinely winter over, no matter what humans do to the environment. However, a surveillance program was set up with the CDC and the New York State Health Department. The following is the sequence of events and the public health policy response:

In July 2000, Dr. Antonia Novello announced that a fast and sensitive DNA PCR test showed West Nile Virus in mosquitoes from Westchester and Suffolk counties, the center of the previous year's outbreak. Both Aedes japonicus and mixed Culex pools of mosquitoes were carrying the virus. Also two crows from Staten Island and two from Rockland County tested positive, bringing to 19 the number of birds that had tested positive for 2000. There were still no human cases. Due to this evidence, Dr. Novello contacted the CDC to work on a multi-state, regional plan of action. Health officials were encouraging people to avoid being bitten by mosquitoes, a bit harder now because Aedes was active in daytime and sunlight, and Culex from dusk to dawn. The NYS Health Department website listed ways for individuals to reduce breeding sites of mosquitoes such as ceramic pots, rain gutters, wading pools, bird baths, and low spots in the yard where water accumulated. Between cleaning gutters, drilling holes in recycling containers, installing new screens, changing bird baths every four days, clearing vegetation, cleaning pools, and landscaping, the average New York citizen had a full time job. That job would have to be done as suggested, wearing shoes and socks, long pants and long-sleeved shirt, and mosquito repellent containing DEET (except that most environmentalists who could get through the heat in full gear would balk at using a chemical). The CDC had added that people should stay indoors at dawn, dusk and early evening, and spray clothing (33, 34).

That was the health response in July, when WNV was detected in birds in three New York counties (on Long Island and in Westchester and Rockland counties), bringing to eight the confirmed bird cases in New York, with another case confirmed in New Jersey the previous month. A spokesperson for the Health Department said she did not think that people needed to be alarmed or avoid outdoor activities, but be aware. There were no confirmed cases in humans. Suffolk County had no immediate plans to spray, nor did Westchester County, but officials were using chemicals to kill mosquito eggs in bodies of water (35).

Later in July, two dead crows on Staten Island tested positive for WNV and New York City officials planned to spray with the pesticide Anvil, a substance similar to the natural pyrethrum derived from chrysanthemums. Concerns about spraying still outbalanced worry about a new virus in the public mind, although if this virus had received the public attention that Ebola virus did, the balance might have swayed the other way. It was only too late in the policy game that the New York Times reported the human effects of WNV and its outbreaks in Israel, South Africa and Romania, where hundreds and sometimes thousands of people became ill and the fatality rate was 5–10%. I have not yet seen an encephalitis morbidity rate, only mortality rates. Of course, courts had to approve New York City's spraying. But this was made easier when a New York Philharmonic concert in Central Park had to be cancelled because mosquitoes there were found to have WNV. The New York City mayor urged citizens not to panic, but also had no plans for aerial spraying, relying on ground spraying, and only in those burroughs where infected birds or other evidence of WNV was obtained. Aerial spraying had been much criticized the year before, as had the use of malathion. Officials still expected only sporadic cases (36)

At the end of July, WNV was found in a dead crow in a Syracuse suburb. A county public health director gave the response that people be aware and take precautions, including wearing clothes that covered their bodies and getting rid of stagnant water. By now, the official public policy response included (and was primarily) the "Fight-the-Bite Campaign," public relations spots in the media. Surveillance (which was working) and public education (which could accomplish exactly what?) formed the backbone of the policy. At the end of July, the city of Boston also joined the list with crows indicating the presence of WNV, and where officials contemplated spraying. A 12-year-old child was hospitalized with encephalitis but test results were not public, and the news reported that so far there were no human cases in Massachusetts. No state-wide plan of spraying, however, was agreed upon. Two dead birds with WNV found 40 miles apart meant that the virus had likely penetrated the state. Still, health officials continued to hope they could get a jump on the virus (37, 38).

In August, a Rochester paper was still reporting that western New York had no reported cases of WNV, although five more counties had confirmed its presence (Albany, Warren, Washington, Dutchess, and Putnam). The paper reported no "confirmed" human case. By now, the NYS Health Commissioner was acknowledging that the virus was likely present throughout the state. Dr. Novello's policy response was to advise individuals to reduce stagnant water and protect themselves from being bitten, a very tall task if one thinks reasonably about it. She didn't feel the news should result in unnecessary alarm. And state officials said the threat was not great enough upstate to begin a spraying program such as the metropolitan areas were using. Instead, the public serrvice announcement of "Fight-the-Bite" was running on radio stations statewide (39).

A dead crow infected with WNV was found in Columbia, Connecticut, but there were no plans to spray the town. That state's chief entemologist predicted the virus would spread across the state and New England. By this time, New York City had ruled out 160 possible human cases and was waiting for test results on 40 others. In New Jersey, officials had referred six suspect cases to the CDC and were awaiting test results (40).

But in the first week of August, the New York Times reported that a 78-year-old man from Staten Island had WNV encephalitis. The man had recovered, as, the news story said, the majority of infected people did. There was no news on the other cases tested for WNV, and at least one confirmed case in upstate New York was not being made public. The news was also suggesting that WNV will not pose a major threat to public health in the U.S. Meanwhile, on the Canadian border, "sentinel chickens" were stationed to sound the first alarm that WNV had spread to Canada (41, 42).

By the end of August, our local county health director wanted us to think WNV was already in our county (which it was), because 29 counties in New York, half the state, had documented the presence of WNV. The official said the map was getting denser and denser. The health director felt the WNV problem would only get worse if not taken care of now and that the county was considering using larvicides but not spraying. In a burst of optimisim, he said: "We really feel this is a problem that will be with us for the next several years." It is likely, of course, that WNV has enough of a foothold to be with us indefinitely into the future, endemic. The type of spraying being done would not reduce the total number of mosquitoes with the virus to any sufficient degree, but the larvicide contained a bacteria to kill the larvae before they hatched. This, if applied adeqautely, would have an effect on mosquito populations, much more effective than urging landowners to eliminate standing water on their property, another health department suggestion. Meanwhile, still claiming no documented case of WNV in birds, our county was waiting for test results on 20 dead crows from the county, but not expecting the results for several weeks. When citizens reported dead birds to the health department phone number for such reporting, they were told that the county was now only taking the count of dead birds, no longer picking them up for testing, but that was not reported (43). The reality of implementing paper public policy was not made clear to the public. In fact, there was only one laboratory in all of New York State testing dead birds for WNV, run by an excellent scientist, Ward Stone, but severely overwhelmed by the task. No new staffing or facilities were provided, no new distribution of economic resources. For all the talk about surveillance in the plans, in the real world, it was not possible to do adequate surveillance.

By the end of August, the New York Times reported that an 87-year-old Brooklyn woman was in critical condition with West Nile Fever, and the city was planning to spray the Bronx, Brooklyn, and Staten Island (44).

By the end of September, the understaffed State lab finally got results to our county. Dead crows, some falling from the sky, some on soccer fields, were found in all areas of the county with growing numbers, and hawks, bluejays and robins were also infected. One local hospital emergency room was doing spinal taps on anyone who came in with severe headache and fever and its lab was getting backed up. The country sent a brochure to every mailing address with hints on fighting WNV ("Protecting Yourself"), including bulldozing property to remove standing water areas (45). Israel was reporting a serious outbreak of the same virulent strain of WNV, with 180 ill and 13 dead. Israel suggested replacing the sentinel chickens with geese, since their geese flocks were being decimated and geese appeared even more sensitive than chickens. In Jerusalem, a rabbi was seeking divine intervention, reciting a prayer which Jewish tradition says dates back to Moses and wards off the Angel of Death. Extermination teams were spraying against mosquitoes.

What are the limitations of our knowledge of this infectious disease? We do not know what conditions generate the spread, according to Dr. Vincent Deubel of the Pasteur Institute who has studied WNV outbreaks for 15 years. Robert McLean, a government biologist and an expert in WNV and birds, concluded the situation was beyond containment and that the population of North America would have to live with the virus as best it could. We do know, too late, that WNV arrived in New York City, likely Queens based on tests of house sparrows (the crow population in Queens fell 90%). We know that migrating birds have first taken the virus northward, and recently the route seems both northerly and southerly. We know the Israeli strain that came to New York is more deadly to birds than the typical strains. We have a top-ten list of possible carriers of the virus to New York from Israel being worked on by Dr. Martin Cetron of the CDC: the million live birds (mostly poultry) imported through air shipments to New York, imported horses, imported amphibians, hitchhiking mosquitoes, or human travellers from Israel whose immune systems were compromised (which could include HIV-positive persons, persons being treated for cancer with immunosuppressant drugs, and persons with organ transplants using immunosuppressant drugs), for example (46).

What we don't know is whether human illness from WNV will be a sporadic event, or whether this strain of the virus can amplify, especially since this strain has not followed the expectations of medical experts, has already surprised and may surprise again. Experts do feel that even with surveillance and spraying, it will be likely that when WNV moves into the human population it will come as a surprise. Since WNV is in a new environment, we can't predict if it will follow old patterns. We know this strain is more virulent in birds but don't know if it will be in humans if it reaches critical mass and moves from sporadic to endemic, as it has in other countries. The media spotlight might then fade, but WNV will then become one of the diseases in North America for which a differential diagnosis will have to be done (47).

What can we learn so far from WNV in North America and the public health policy response?

The surveillance component of public health was put into effect fairly rapidly, although the public was not always given the most current information, and labs doing the testing were backlogged. Some county health departments restricted themselves to doing dead bird counts. Surveillance continued watch, as the virus progressed from the first small area of outbreak, where containment and elimination might have been possible, to the second year where the outbreak was statewide and region-wide and no longer possible to contain or eliminate.

Responses to the surveillance alert were not adequate. Localized spraying too late in the season lost the easiest chance to stop the virus from gaining hold in the first year. Public health experts tried to find a balance between the risk of WNV and the not-well scientifically documented risk of pesticide spraying. The public, however, had made up its mind that insecticides were high-risk, even though information sheets from the CDC and NYS Department of Health tried to show that there was little risk. Because of this ideological rather than scientific pressure, officials did not spray as aggressively as could have been done. Instead of a classic public health response to surveillance data, a decision to put most of the burden on the individual through public education was made and implemented in New York State, where the outbreak had started. A public relations campaign, "Fight-the-Bite" was promoted. But as biology is not political ideology, it is also not media public relations. The campaign might give the impression that something substantial was being done, but the reality (not the conversation) was that there was no measure taken to prevent West Nile Virus from becoming endemic in North America. The average citizen has always hoped to avoid being bitten by mosquitoes, but has always failed in that hope. Nor can the average citizen rearrange the ecosytem sufficiently that mosquitoes will not have a place to breed. This public relations postmodern conversation did nothing but appeal to those experts who believe a media spot solves problems.

The focus appeared to be on the Don't Panic Scenario, minimizing the potential consequences of a new virus that can cause encephalitis and is lethal to some wildlife. The public was given information, but it was skewed in ways that would reassure and make the case for accepting the presence of a new pathogen on this continent. Full information, however, was not rapidly given. The reality is that public health measures were not taken to give some chance of preventing an insect-vectored disease with 5–10% mortality and unrevealed morbidity from obtaining first a foothold on the continent and then a regional presence.

THE ETHICS OF PUBLIC HEALTH POLICY

In the body of this chapter I've noted seven ethical concepts important to public policy and health policy expertise/planning.

1. *Conflicts of Interest.* Because different levels of a system have varying characteristics, needs and interests, it is very likely that any policy making will contain naturally embedded conflicts among these levels. Policy has to be made within a conflict-of-interest potential. Ethically, the experts need to be aware themselves of what their conflicting interests are, and they need to communicate this to the public. Operators of blood banks will want to consider the best-case scenario (Don't Panic) as the accurate one, and unless they can correct for this bias, will make wrong evaluations and choices all along the process of policy planning and implementation. Individuals who need the availability of a blood transfusion but also wish that supply to be safe will be inclined to want zero risk of their health and life, and will consider the worst-case scenario (Doomsday) more easily. Virologists who need to fund their research labs will have to be aware of their need to see emerging and resurging pathogens as extremely important public health issues approaching an Andromeda Strain risk, while funding agency directors trying to allocate limited resources will hope they can gamble on inaction. In itself, there is

nothing unethical about conflicting needs and interests. This is a statement about natural systems in the real world. It is when the players are not aware, or cynically do not make others aware, of these conflicts that an ethical problem arises. It is compounded by bureaucracies' needs to insulate from blame and responsibility, or to not admit problems.

2. *How and Where on the Don't Panic Scenario vs. Doomsday Scenario to Place the Problem.* Any point along the continuum of risk response is a possibility, but some points make more scientific and ethical sense than others in a given situation. In addition to factual ambiguity, there can be ideological assumptions, as readily seen in the apocalyptic placements of global warming, destruction of the rain forest, eliminating biodiversity, or defining scarce resources. The role the expert is playing in the policy planning can also be an important factor, for example, a health director trying to avoid the political consequences of an unpopular policy like spraying. AIDS/HIV is an interesting example of the presence of both contradictory scenarios in a consensus policy. The Doomsday Scenario was used when funding was an issue. The Don't Panic Scenario was used when restrictions of Gay Rights was an issue. The Doomsday Scenario was used in the AIDS Quilt Name Project, to produce dying heroes that humanized homosexuality for the public. The Don't Panic Scenario was used in the struggle to keep Ryan White in public school in spite of being HIV positive, and in spite of some low risk of casual transmission. The ethical obligation of the expert here is to keep within scientific probabilities and reasoned contexts, and create a valid scenario. We know, for example, that the public did not panic so badly over the great pneumonic plagues of Eurasia that civilization came to an end or those in power were slaughtered. We know that it is very difficult for any pathogen to completely wipe out a species, and though the American elm tree is almost a counter-example, it is a rare example. False reassurances destroy the trust on which public policy is based by creating doubt in the public mind about reasonable safety. Crying wolf too many times also creates doubt in the public mind and dismissal of all risks, even real ones.

3. *Informing and Involving the Public vs. Causing Harm by Premature or Unjustified Precautionary Actions*. There is often a lack of real, substantive public involvement in policy making, as expressed in the term for such experts, "wonks." It is easy, but unethical, to dismiss the public as prone to panic, and to reserve real analysis and policy for such experts. The British Mad Cow example illustrates how a select committee kept information from the public that should have been available, because, I assume, of concerns that the public

would be unnecessarily frightened at that stage of inquiry, and that the economic damage to the cattle industry would be unjustified. Again, the issue of trust in public policy is vital here. Without it, such policy is difficult to implement, and censoring information by making a two-tier system, one for the common man and one for the expert, is self-defeating as well as unethical. Sometimes censorship or secrecy is hidden by a pseudo-involvement strategy. The forms of involvement are present, the trappings of an interaction are displayed, but the reality is far different. Anyone who has attended a department of transportation hearing about road-widening or changes knows the engineering decisions have been made, token public representatives have had their names used, and after all plans are in place, a public meeting is called to allow empty talk that does not impact on the decision. Everyone, having gone through the paces of what appears to be participatory democracy, goes home to the reality of the finalized plans. There may be some situations where the harm is severe enough that not informing or involving the public would be ethical. Wartime situations qualify, although the secrecy acts are often abused. My argument would be that these situations are few and far between, and that informing and involving the public in public decisions is a required balance to the power of experts and the strategems of public relations.

4. *Expert Standards of Objective, Critical Thinking/Expert Obligation*. Respect and trust are earned, and experts can be respected and trusted if the following standards are maintained. An expert ought to correct for bias and conflict of interest to the best of his or her ability, and should identify the position so that others may help in the correction. An expert ought to be open to new experiences or data, following the inductive, scientific method, and not prematurely close analysis or ever fully close analysis. An expert should be honest with public information, not just in terms of direct lies and falsehoods, but more importantly in terms of deliberate, subtle deception. An expert should avoid public relations techniques which are manipulative and do not create ethical policy. Finally, an expert has the obligation to be truely competent in his or her field. Experts will disagree, which is not a bad thing given the openness of science.

5. *Risk Assessment and Risk/Benefit Analysis*. Risk assessment and anlysis requires acceptance of trade-offs as part of the nature of systems. First, the assessment needs to be as verifiable as possible. It needs to be broken down into its major components and those components needs to be adequately understood by all involved. The first component, the risk probability figure, is a statistical figure that needs some fleshing out as a concept for the public. It

is basically a gambling figure, an idea of the odds a gambler would want to know before risking a bet. Ethics sometimes errs too far on the side of prudence, however. Taking risk is frequently the ethical thing to do and should not be seen in a negative light. It is important to have this figure as accurate as possible, and not some assumed number. There are so many one-in-a-million risks that are nothing more than nonsense that it is rare to find an actual one-in-a-million risk that is scientifically accurate. The second component is the severity of the consequences if one loses the bet with nature. The odds of our local nuclear power plant having a Chernobyl-type event are very, very low, but people are uneasy about nuclear power. It is not an entirely irrational fear, since the results of such an event would be severe (though not as severe as the Doomsday Scenario would suggest). Because of the severity, people are inclined to restrict their gambling in such cases. The third component is control over the choice of taking the risk. That might be best illustrated by a conversation I once had with the quality assurance director of that same nuclear power plant. He was a German immigrant who had been in the Luftwaffe during World War II. His crew was going up to do training exercises one day, but he had just lit a cigarette and didn't want to waste it. He told them he would catch the next exercise, and remained on the ground. Their plane crashed with all lives on board lost. Smoking, that public media high-risk behavior, had saved his life. He was in control of whether to risk lung cancer or not by smoking, and he was comfortable having that control. His risk assessment before the crash reflected that comfort with having a choice. After the crash, over which risk he did not have control, he comforted himself with the idea that every day on earth that he now lived was a bonus, an extra day (a fatalism). He continued to smoke throughout his life, no longer concerned with any risks. His behavior also illustrates the ethical, evaluative nature of risk/benefit analysis. We can conflate risk with cost, we can look at a person's willingness to pay certain prices as a marker, but in the end, risk/benefit analysis is an ethical judgment made within an ethical theory, not a political or economic science, or even a statistical trade-off theory.

6. *Uncertainty Factor.* Clinicans have always had to accept accountability and uncertainty at the same time. This is very difficult, of course, but a reality. Policy planning will contain an uncertainty factor that may be small or may be substantial. As with a clinician, uncertainty is not an excuse for negligence, incompetence, malevolence, or deception and manipulation. Remaining open to feedback, as the classic clinician did, rather than becoming overly-committed to protocols, hearing the ping of experiences that do not fit the carefully-constructed but still only an approximation analysis and plan, is both

an ethical and scientific requirement. As William James put it, one needs always to hear the cries of the wounded, those wounded by evaluations, plans, strategies, and judgments (48).

7. *Balance in a System*. Finally, the core ethical concept, what makes systems bioethics a useful tool for policy as well as ethics, is a balance point within a natural system. It can be quite complicated. In AIDS/HIV, for example, the needs of community/global health, public health, and persons with HIV are often in conflict but also closely interconnected. HIV-positive persons with compromised immune systems can be reservoirs for emerging, resurging and resistant pathogens such as tuberculosis or perhaps West Nile Fever. This threat requires balancing human needs for free thought and action with human needs for protection against pandemics, but too much sacrifice of any level of a natural system can bring down the system. Balance has to be the flexible integration of as many functions at all levels as is realistic. Balance is also not static, but controlled state oscillation. Some preference can ethically be given to critical malfunction solutions at one level, but only if feedback from other levels is considered and acceptable. Only in extreme cases would extreme ethical choices be made that sacrifice major needs of any one level. Finding this balance is not Aristotle's Golden Mean, though that philosopher was likely glimpsing some of the nature of systems when he constructed his ethical theory. The ethical choice is not automatically the middle, but a balance that may shift as situations shift, that may sometimes require the extreme sacrifice, that often requires the ongoing compensation for multiple reactions at multiple levels. None of these, however, can be governed by ideology, which cannot supply the dynamic concepts that systems theory requires.

In infectious diseases, it pays to consider the ancient history of plagues, both epidemic and pandemic. Human exploration and population movements are basic to human social behaviors, and have always involved risks. Risk is often an ethical behavior, and prudence is not always a virtue. The risks, however, can be high. The Portuguese sailors of the 16th century knew that:

> "Beware and take care of the Bight of Benin,
> There's one comes out for forty goes in."

REFERENCES

1. Landau-Stanton, J., & Clements, C. D. (1993). *AIDS, Health and Mental Health: A Primary Sourcebook*. New York: Brunner/Mazel.
2. Clements, C. D. (1994). Systems bioethics, AIDS, reproduction, and the pandemic destroying population groups. In: E. Bittar & N. Bittar (Eds), *Principles of Medical Biology*, Vol. 1A (pp. 88–114). Stamford, CT: JAI Press.
3. Shilts, R. (1987). *And the Band Played On: Politics, People and the AIDS Epidemic*. New York: St. Martin's Press.
4. Shilts, R. (1987). *And the Band Played On: Politics, People and the AIDS Epidemic*. New York: St. Martin's Press.
5. Landau-Stanton, J., & Clements, C. D. (1993). *AIDS, Health and Mental Health: A Primary Sourcebook*. New York: Brunner/Mazel.
6. Shilts, R. (1987). *And the Band Played On: Politics, People and the AIDS Epidemic*. New York: St. Martin's Press.
7. Landau-Stanton, J., & Clements, C. D. (1993). *AIDS, Health and Mental Health: A Primary Sourcebook*. New York: Brunner/Mazel.
8. Clements, C. D. (1999). Systems bioethics: The future of epidemics and medical ethics. In: R. B. Edwards & E. E. Bittar (Eds), *Advances in Bioethics* (pp. 31–60). Stamford, CT: JAI Press.
9. Palca, J. (1991). The sobering geography of AIDS. *Science*, *252*, 372–373.
10. Clements, C. D. (1994) Doing the right thing: Systems ethics and AIDS. In: J. Landau-Stanton & C. D. Clements (Eds), *AIDS, Health and Mental Health: A Primary Sourcebook* (pp. 300–327). New York: Brunner/Mazel.
11. The Associated Press. (9/16/2000). *Mad Cow Disease passed by blood, research finds*. New York Times.com.
12. Goldberg, C. (7/18/2000). *U.S. will seize and destroy sheep at risk of an infection*. New York Times.com/library/national/science.
13. Blakeslee, S. (5/23/2000). *Clues to Mad Cow Disease emerge in study of mutant proteins*. New York Times.com/library/national/science.
14. Blakeslee, S. (5/23/2000). *Clues to Mad Cow Disease emerge in study of mutant proteins*. New York Times.com/library/national/science.
15. Daley, S. (5/7/2000). *Rising rate of Mad Cow Disease alarms Europe*. New York Times.com/ library/world/europe.
16. Blakeslee, S. (5/23/2000). *Clues to Mad Cow Disease emerge in study of mutant proteins*. New York Times.com/library/national/science.
17. Daley, S. (5/7/2000). *Rising rate of Mad Cow Disease alarms Europe*. New York Times.com/ library/world/europe.
18. Sagoff, M. (1988). *The Economy of the Earth*. Cambridge, England: Cambridge University Press.
19. Landau-Stanton, J., & Clements, C. D. (1993). *AIDS, Health and Mental Health: A Primary Sourcebook*. New York: Brunner/Mazel.
20. Caton, H. (1994). *The AIDS Mirage*. Sydney, Australia: University of New South Wales Press.
21. Fumento, M. (1990). *The Myth of Heterosexual AIDS*. New York: Basic Books.
22. Shilts, R. (1987). *And the Band Played On: Politics, People and the AIDS Epidemic*. New York: St. Martin's Press.

23. Clements, C. D.(1994). Systems bioethics:AIDS, reproduction, and the pandemic destroying population groups. In: E. Bittar & N. Bittar (Eds), *Principles of Medical Biology*, Vol. 1A, (pp. 88–114). Stamford, CT: JAI Press.
24. WHO Global Programme on AIDS. (1/4/1993). *The current global situation on the HIV/AIDS pandemic*. Geneva, Switzerland.
25. Brandt, A. M. (1988). *The syphilis epidemic and its relation to AIDS. Science*, *239*, 375–380.
26. Grmek, M. D.(1990). *History of AIDS: Emergence and Origin of a Modern Pandemic* (Translated by Maulitz, R. C. and Duffin, J.) Princeton, NJ: Princeton University Press.
27. Morse, S. (1993). *Emerging Viruses*. New York: Oxford University Press.
28. Preston, R. (1994). *The Hot Zone*. New York: Anchor/Doubleday.
29. McGinn, B. (1994). *Antichrist: Two Thousand Years of the Human Fascination with Evil*. San Francisco: Harper.
30. Centers for Disease Control and Prevention. (6/21/2000). Division of Vector-Borne Infectious Diseases: West Nile Virus. Cdc.gov.
31. New York State Department of Health. (5/19/2000). *West Nile Virus (WNV) Update: Veterinarians*. Health.state.ny.us/nysdoh/westnile.
32. Centers for Disease Control and Prevention. (6/21/2000). Division of Vector-Borne Infectious Diseases: West Nile Virus. Cdc.gov.
33. Centers for Disease Control and Prevention. (6/21/2000). Division of Vector-Borne Infectious Disease: West Nile Virus. Cdc.gov.
34. New York State Department of Health. (5/19/2000). *West Nile Virus: Frequently Asked Questions*. Health.state.ny.us/nysdoh/westnile/education.
35. Kershaw, S. (7/9/2000). *West Nile Virus detected in 3 New York counties*. New York Times.com/library/national/regional.
36. Revkin, A. (8/8/2000). *Clues to an alien virus*. New York Times.com/library/national/science.
37. Noble, C. (7/25/2000). *Dead crows fuel West Nile concern*. Absnews.go.com.
38. Goodman, J. (7/30/2000). *Nile virus not a concern yet. Health officials remain vigilant, however, after infected bird is found in Syracuse*. Rochesternews.com.
39. Rivoli, M. D. (8/4/2000). *West Nile Virus could be statewide*. New York: Democrat and Chronicle, Rochester.
40. Koh, E. L. (8/4/2000). *West Nile found in dead birds in five more upstate counties*. New York Times.com/library/national/regional.
41. Revkin, A. (8/8/2000). *Clues to an alien virus*. New York Times.com/library/national/sciences.
42. Murray, T. (8/8/2000). *West Nile Virus in U.S. found in birds with migratory patterns here. The Medical Post* 36(27):5, Toronto.
43. Larvacides, other steps may be taken against West Nile. (8/25/2000). Rochestertoday.com/news.
44. Ireland, C., & Livadas, G. (8/30/2000). *West Nile hits Monroe county*. Rochester, NY: Democrat and Chronicle.
45. Doyle, J., & Doniger, A. S. (9/2000). *Mosquitoes and West Nile Virus: Protect Yourself*. Rochester, NY: Monroe County Department of Health.
46. Revkin, A. (8/8/2000). *Clues to an alien virus*. New York Times.com/library/national/science.
47. Revkin, A. (8/8/2000). *Clues to an alien virus*. New York Times.com/library/national/science.

48. James, W. (1980). The moral philosopher and the moral life. In: L. E. Navia & E. Kelly (Eds), *Ethics and the Search for Values* (pp. 369–390). Buffalo, NY: Prometheus Books.

6. ABORTION: HUMAN WORTH, THE END OF ETHICS AND POLITICAL ENTITLEMENT

HOW POLITICAL IDEOLOGY HAS REPLACED ETHICS IN THE ABORTION QUESTION

One of my columns for *The Medical Post* described some of the cultural implications of the absolute position on the civil right to privacy, and it generated a letter from a physician who appears to be an abortion-rights activist that in a few words makes the point of this book (1). Dr. Mark Kreiger's opinion piece was entitled, appropriately enough, "Abortion is about rights, not ethics." He was responding to a column of mine that analyzed the risks of a pro-choice position on abortion, forgetting that I also argue for a limited pro-abortion position in ethics. For him, the column was yet another attack on a woman's absolute freedom of choice and control of reproduction. The sentence that struck me as encapsulating the argument that bioethics is actually political, postmodernism was: "Abortion and women's rights go hand in hand and the history of abortion has been much more a matter of politics and the enforcement of the dictates of male-dominated legislatures and religious bodies than of ethics." For the "patriarchy" of legislatures and religious institutions, he substitutes the neoleft political ideology of postmodernism, the absolute privacy rights of women during pregnancy. He means, as does postmodern bioethics mean, that such bioethics issues are and should be political issues, resolved through political tools of power and strategy. Content ethics must be disregarded, and in fact is impossible in such a framework, as my column had suggested was the case.

The ideology is quite clear. Political rights are paramount, at least for certain entities that are assumed to deserve them. Women, in keeping with the current feminist assumptions of postmodernism, must have an absolute privacy right because of the general women's rights activism within this ideology. Men, on the other hand, are most often dominating oppressors whose power must be broken. A kinship relationship, patriarchy, is converted into a political slogan representing women as always victims and men as always oppressors because of the structure of society and family. Theological content-ethics is seen as authoritarian and oppressive, requiring for its decline the "increase in the legislative voice of women and their supporters," again, a political strategy of power. Further in the piece, the assumed economic agenda also surfaces, along with the internally-contradictory preference for centralized state goals. The argument is made that it is not in society's interest to insist a woman continue the pregnancy. The women will be young, single, in school or poorly-educated, unable to pursue education or employment if pregnant, and incapable of getting their lives in order. That will result in a heavy financial burden for the state, with welfare and social assistance costs, the argument assumes. Even if willing to accept these costs, most who make this argument do not feel taxpayers should bear the costs of children raised in socio-economically-deprived situations, adding the assumption that a healthy and happy childhood requires an upper-middle-class income. This last assumption, though it is inconsistent with the neoleft perspective at first glance, must be seen in context. Consistency is not required by postmodernism, so it has never been seen as a problem for postmodern bioethics. All conversations that imbibe the political consensus are equal. The activists making this assumption happen to come from the higher strata of society in general and also have a faith in the "health belief model" which sacrifices individual decisions and lifestyles for a state-approved lifestyle. I will analyze the health belief model in Chapter 8, and here only want to list it as one of the politicized components of current bioethics.

The central point, unfortunate for the possibility of ethics within postmodern bioethics, is that deeply-ethical questions such as membership in an ethical community or the ethical status of an entity are dismissed in favor of a political strategy to implement certain agreed-upon assumptions that are political in character. Ethics has been reduced to politics.

A letter written by Marilyn Wilson, executive director of the Canadian Abortion Rights Action League, about another of my columns on abortion reflects the same position as Dr. Krieger, and demonstrates he is not atypical (2). Wilson was concerned that my comments would contribute to the nasty cultural war surrounding abortion, but her concern was likely generated as a fear that the comments would contribute to the strength of the other side in this war in which she wished to declare victory. Again, the argument centered around civil or political rights, stating we must "accord women the same freedom from ownership . . . that we accord every member of our culture." Her description of these rights, however, was in absolute terms that have not been recognized by any court with the exception of the latest U.S. Supreme Court ruling on state limitation of late-term abortions. None of us enjoy such an absolute right of power to decide what to do with our bodies as this letter implies for pregnant women. There is an even greater power given to the female portion of our society, however, by this form of stating the problem. Only women are given an almost absolute power to determine who is and who is not a member of the ethical community, or to disregard the question completely in their decision and its justification. In fact, the decision needs no justification, it is argued. Wilson argues that women do have the right, if they wished it, to determine who is a member because women are valuable persons. The logic of that would escape all but a postmodernist who did not care about logical standards of reasoning. We are all valuable persons intrinsically, but I would argue we all do not have nor should have the individual, arbitrary power to decide who else is a member of the ethical community. That decision is part of the primary justification for the existence of the state, the police power or commonwealth power so easily called on by bioethicists in other questions, but refused significance in the abortion question. It does not seem to me possible to have a viable state when individuals can make the determination of who is an ethical entity in that state. We do have to achieve social consensus on the status of embryos and fetuses, and that consensus should not be a forced one through the use of political power alone. We are not likely to achieve that consensus within postmodern bioethics since it does not have the tools for the rational discussion required. The headline of the letter makes clear the postmodern ideology: "Abortion is about a woman's right to control her body."

Even with a letter writer who did realize that Dr. Krieger's discussion "ignored the ethics debate", Dr. Thomas Millar, the culturally-conditioned political framework was hard to put aside (3). Dr. Millar wrote that ethical issues involve rights that are in conflict, the right of a woman to control her reproduction and the right of the unborn child to develop into a baby. In stating the problem in this way, he has used two concepts of "rights" that come from very different ethical theories, the concept of a civil or legal right and the concept of a natural or human right. In our current culture, however, the favored meaning of "rights" is the former, since it is a political definition. He does see, however, even in the rights-language, the really core ethical issue. He says if there were a way to define women as non-persons (in other words, having no ethical standing), the debate would be finished. If there were a way to define the fetus as a non-person (no ethical status), the debate would also be finished. He then assumes that both women and fetuses are persons, and that this debate will not end. His letter is a plea for having a full-scale debate on the real ethical issue, what entities have membership in the human ethical community, and he understands that the pro-choice position of avoiding the debate by focusing on the political right of the woman to choose is an "evasion" and wishful thinking that the debate will end by default or political coercion. His plea is sad: "That is why . . . the issue doesn't go away. Nor will it, until the ethical debate you would deny us is wholly engaged upon, and some solution morally acceptable to the whole society is worked out."

I have suggested the major problem with the position of current bioethics, but in fairness I must also address the problems of the opposition. Although these problems are serious, they do not imply the end of ethics or the total society, and in some ways, though I disagree, I am less fearful of the consequences for humanity of this position. That same column I wrote which generated Dr. Krieger's response also generated a response from Dr. Trevor Watson (4). Dr. Watson did not appreciate that I had come to my limited senses after logical peregrination that "extermination of the unborn can – only on occasion, mind you – be harmful to society. She seems to have . . . stumbled upon the truth that killing human beings is liable to have an undesirable consequence. What an insight." Here we have the ethical status of the embryo or fetus asked and answered, absolutely, without the possibility of further analysis. It is a full-fledged member of the ethical community, a human being. I have always philosophically argued that killing a human being has harmful social consequences, although similar to the argument for a just war, I have concluded that it can sometimes be justified ethically, sometimes required ethically. Violence is, more often than we wish to think, an ethical action.

Watson is aware that I have a content theory of ethics, because he quotes my use of the terms "good" and "evil" and my affirmation of "reason and science." He understood that I argued for an ethical system based on the nature of man in the real world, and he was upset that his profession was making decisions according to the "professional and community mores of the moment." He understood the importance of the value of life, especially for physicians. However, he wanted the conclusion of the Judeo-Christian ethical system (which is much more Roman than either Hebrew or Christian) to be accepted. He recalled that this system had come down through thousands of years and trials and errors, and David Hume might agree with him that historical experience should count for something, but not everything (5). He quoted G. K. Chesterton's defense of orthodoxy as the only safe guardian of morality, order, liberty, innovation and advance. Unfortunately a systems theory analysis would find that claim untenable.

Unlike postmodern bioethics, this position does fundamentally consider the ethical status of the entity and does it with a philosophically-deep literature. The conclusion, however, is not foregone. It is not so immediately clear what the ethical status of a human egg or sperm are, or human DNA for that matter, nor is it transparently obvious what the status of an embryo, nonviable fetus, or viable fetus are. What does seem reasonably clear to me is that we cannot allow this ethical question to be answered politically. Politics cannot deal with man's place in the universe. It cannot give a definition of man. It does not have love (for self, for family, for community, agape) in its concepts. It provides a meager purpose for human existence, power.

What had the column said that placed it outside both the pro-choice and the pro-life positions? The column had analyzed the sad case of a 12 year-old girl who had been impregnated by, in all probability, her brother. Although she had been taken for medical examinations, the sexual act and the pregnancy were not discovered until the fetus was viable. I had argued for the complexity of the case and for at least the sophisticion in ethics to avoid absolute ethical principles, on this in agreement with Joseph Fletcher and Richard Taylor (6, 7). I contended the question could not be answered in either the pro-choice or pro-life framework. Instead, I argued that there was a need to develop ethical standards rather than political ones, and not fool ourselves that we could live in the ethical vacuum of individual choice. Consequences were important in bioethics, and the end did partially justify the means, but only partially. The means also partially determined the end. The issues were clear: the ethical response to brother-sister incest, the ethical status of a viable fetus, ethical concerns about pregnancy in a child, legal concerns about statutory rape and late-term abortions, who would ethically be the decision maker, and the ethical obligation to the physical and psychological health of the girl.

Because of state law, the parents petitioned the courts to allow them to take their daughter to Kansas where an abortion could legally be done, but that is a legal answer, not an ethical one. What good consequences could be expected from the abortion of a viable fetus? In terms of the girl's health, there would be little difference between such a late abortion and the natural delivery of a baby in a hospital setting. The abortion shortened the pregnancy, but very little, saving the girl a few weeks of pregnancy. Both options had negative psychological effects. Going through delivery and giving the baby up for adoption can be a long-term trauma. Keeping the baby could create family and personal problems, because the father is also the brother and because the baby would be unfairly stigmatized as the product of incest. Killing a viable fetus also has harmful psychological consequences, since no killing is an easy thing to justify personally, even for war veterans. Biologically-primed to become a mother, there will be serious issues of loss for the girl, and these may not be addressed. Preventing a genetic disease because of the incest is a statistical game and depends on the genome of the parents of the child. Not all incest results in a genetic disease, although the risk is increased statistically.

A consequential analysis does not give good reasons for terminating a viable fetus. The only other justification would be the choice of the child or parents. With a viable fetus, we would hope for better ethical justification. There would need to be considerable benefits to the girl, and serious harm to her if the pregnancy were not terminated. There could be serious legal harm to the brother if it could be determined that he had intercourse with his sister, and that determination could have been made in the home state through DNA analysis of the baby when delivered. His ethical standing, however, has been compromised by the incest.

The deontological argument that human life must never be terminated, and that a viable fetus has the same standing as fully-developed human beings would still require considering the damage to the girl as well as the harm to the fetus. Because the pregnancy was so far along, the damage had already been done, basically. Aborting would not save her from the harm to herself. This is an unfortunate case where unacceptable harm to the girl has already been done. We cannot prevent that harm, and will now do harm to the fetus through aborting. If it can be established that a viable human fetus does have the same status as other persons (and the law recognizes, except in abortion, the legal standing of a viable fetus through much case law on drug use in pregnancy, auto accidents, and battery of a pregnant woman), then there needs to be very powerful arguments to justify killing that fetus. Choice, convenience, the risks of incest would not be powerful enough.

The political argument from choice is flawed and dangerous. Society can't allow individuals to determine for themselves alone membership in an ethical community. There is no justification for such power and such anarchy is not accepted even within the postmodern ideology on other bioethics questions, because it runs contrary to another assumption of that ideology, the centralized and total society. We can't survive with ethical anarchy because we are a social species, because our nature requires doing ethical evaluations. Pro-choice destroys the possibility of ethics, particularly in the abortion question, but finally for all human activity. Pro-life absolute principles destroy the possibility of understanding the human being and specific interactions with the natural world, of ethics being for humans. However, general ethical rules can be constructed despite arguments to the contrary from postmodernists in bioethics. General rules will have exceptions. The case of this young girl, however, does not seem to be that exception to the general rule of the value of human beings. Her pregnancy is too far advanced and little good could come from the abortion. There are reasonable arguments that a viable fetus does have ethical standing, which can only be overridden for serious reasons such as demonstrated, severe harm to the pregnant woman. There are harmful consequences for the abortion option that will create evil ripples in the social pool and would be hard to manage. For example, easily terminating a viable fetus and claiming free choice to do that threatens the integrity of our social bonds, of human love, and places us all at risk. Given such high-risk choice without ethical justification, the late-term abortion was unethical. From an ethical view, it was just too late.

HOW MODERN LAW FAILS MODERN SOCIETY

There are times when one lost life is given rather than losing two lives. Effecting the delivery of an almost-full-term fetus' torso and then collapsing the skull to complete the abortion is not a new medical procedure. When I was pregnant for the first time, I purchased an old obstetrics book at a book sale. It was so old that woodcuts illustrated the pages rather than photographs, but it was thorough for its time. One woodcut remains in my mind, an illustration for a desperate intervention to save a desperate situation. The illustration graphically described the medical instrument that looked like a winged wine bottle opener. There were two moveable metal wings on either side but in the middle, replacing the corkscrew, was a straight, pointed piece of metal. The point was meant to pierce the head of the delivering baby stuck in the birth canal. The lance went through the skull and the side wings closed into it to help collapse the head. This was a sad but ethical decision to save the life of the woman, since the baby's life could no longer be saved. Because I think the ethical argument for this is sufficiently sound, I think we all could look at this illustration without flinching or dissembling reality away. If something is ethical, it will withstand the glances of reality. If it can't withstand that glance, if we must look away, then there are serious questions about the ethics of the act.

On June 28, 2000, the U.S. Supreme Court looked away, and with twisted words and thoughts avoided the test of reality. It is hard to read Justice Stephen Breyer's majority opinion any other way. The majority could not find a word to describe what the Nebraska law under challenge had called "partial birth abortion." The act is a partial delivery of a usually viable fetus, and it is abortion, but a more neutral term would be preferable for some. Physicians have begun calling the procedure a "D & X," those arcane acronyms that confuse laypeople and remove some humanism from medicine. "Dilation and Extraction," however, does not describe this procedure adequately. It is no more just a dilation and extraction than a "D & E" is just a dilation and evacuation, though having had a lengthy description from an obstetrician doing D & Es, I am aware of the realities of the procedure and would not come to a prejudgment on the ethics of the situation. Justice Breyer's opinion did, however, seem to have problems with naming the procedure in question, and even D & X seemed too graphic for him. He referred to the intervention as "D," a move that implies volumes and supports Justice Clarence Thomas' point that the court majority hesitated even to describe the procedure they wished to hold constitutional.

It is not a D. It is the use of a pointed piece of metal to pierce the skull and brain of a 16 weeks to full-term fetus whose legs, torso and shoulders have been delivered by inducing, through the vaginal canal. The collapsed brain draining from the skull leads to skull collapse and the delivery of a dead entity. The purpose of the majority of such procedures is the abortion of a healthy fetus by the choice of a healthy pregnant woman. The majority opinion dismissed the qualification of the American College of Obstetrics that its panel "could identify no circumstances under which the D would be the only option to save the life or preserve the health of the woman." What the majority opinion did was give physicians an implied constitutional right to use any rarely-used method provided it is called treatment, no matter what the ethical status of that method might be. Professional control is an important goal as argued previously, but such an absolute right is extreme. Female circumcision is a procedure sometimes done by physicians, yet a worldwide governmental ban would make ethical and scientific sense. Physician-assisted suicide is a procedure, but the same Supreme Court did not find it a privacy right or give physicians freedom to use it. Postmodern bioethics supported severe limits on electro-convulsive therapy or frontal lobe lobotomy. The U.S. Supreme Court did not give men and women a constitutional right to ECT or find the restrictions undue and substantial. Similar to the ideology of bioethics, there is a terrible logical inconsistency in the Court's reasoning about abortion, now most apparent in this Stenberg vs. Carhart decision (8, 9).

In the majority argument, an undue burden has been transmuted into a substantial obstacle, which is not an equivalent concept. Any burden can be substantial, but not all substantial burdens are undue burdens. Still, based on this interpretation of a previous Casey (10) ruling, Breyer goes on to extend the constitutional right to an absolutism. Any restricting law, he argues, has to contain a broad and general exception for the mother's "health." This makes it impossible to regulate abortion at any point in pregnancy, for whatever reason. Breyer even questions the Nebraska law as furthering the state's interest in protecting human life, arguing that the law regulated the method of abortion, not the saving of the fetus from destruction. This sophistic argument plays with words. The Nebraska law would have made it more difficult to destroy the late-term fetus by regulating a method used to do just that. The goal was to save the life; the means was to regulate the method. He is equally sophistic when he argues that the Nebraska law might be applied to D & E procedures, because a substantial part of the fetus could be outside the uterus before the fetus is dead. From what I know of D & E, the fetus is taken apart in the uterus, with some recognizable pieces then suctioned or pulled out. I doubt that a dismembered organism can be considered alive at the seconds after dismemberment, and this seems a spurious comparison. Justice Thomas disagrees with Breyer, pointing out that the AMA recognizes D & X/partial birth abortion as ethically different, since the fetus of 20 weeks or more is killed while partially outside the uterus. It is the reality of D & X which causes Justice Thomas to raise the word "infanticide," and we should not quickly dismiss it. That does not imply that infanticide is always unethical. It means we should recognize the actuality of what we are talking about in bioethics.

What was the understanding of ethics of those supporting the majority (5 to 4) decision? Janet Benshoot, president of the Center for Reproductive Law and Policy, described the day of the Court decision as a time for "Champagne and shivers." Dr. Frederick Hopkins, an officer of Planned Parenthood, insisted there was no such thing as partial birth abortion because the words aren't in medical texts and don't have a medical definition. The majority Justices thought it was irrational to find any fundamental differences between one procedure or another. This superficial appreciation of the worth of human beings and this lack of humanity are common aspects of postmodern bioethics. It follows from postmodernism's lack of basis for any valuation of human life, and its use of a legal tool that is also losing its base for the worth of a human being (11). Instead, there is a focus on power strategies, language games, and in this case, the victory of champagne. The majority ruling is Byzantine at its best, and Byzantium was the epitome of a politicized society where even the chariot races had political ideology as the motive, the Greens and the Blues.

Contrast this with the understanding of ethics, not politics, shown by the minority opinions. Thomas correctly worries about infanticide, the dehumanization of a viable fetus, and the trivialization of human life. He rejects the majority's sanitized description and insists we look at the reality of the procedure. He wishes to be consistent with other laws that give viable fetuses a moral standing. Justice Kennedy insists in his dissent that there is a moral difference between the D method and other abortion procedures, and praises Nebraska for choosing to ban a method "many decent and civilized people find so abhorrent as to be among the most serious of crimes against human life." He insists the Casey decision (of which he was a part) does give states a constitutional power to promote respect for life. The dissenting opinions were read in court because these justices were aware of the seriousness of a state's inability to defend the lives of its citizens.

The supporters were worried about the closeness of the decision and its implications for the U.S. presidential race. Those in disagreement might find champagne inappropriate, and might shudder rather than shiver at the political take-over of ethics. The probability of a human value system became even less.

The Gallop Organization has been polling on abortion since 1975. Their combined polling data from 1975–1998 showed some trends that fit poorly with the Supreme Court decision. In 1998, 48% of those polled considered themselves pro-choice, down from 56% in 1983. In 1998, 45% considered themselves pro-life, up from 33%. There was an almost even split. In 1998, only 23% felt abortion should be legal under any circumstances, down 10%. Under some circumstances, 59% think abortion should be legal, leaving 41% disagreeing or unconvinced. In 1998, 48% thought abortion was an act of murder, up 8%, with 45% thinking it was not murder. This large percentage opposing abortion and thinking it is murder has been disenfranchised, frustrated, locked out of any discussion about the merits of its philosophical position, denied political expression – in other words, repressed and oppressed. When all means of coming to a social resolution on this important ethical issue are blocked by political power, it is not hard to predict the alternative will be violence of some sort. The political entitlement claimed by pro-choice activists and supported by postmodern bioethics, based on ideology and current power, will not hold the center indefinitely given the Gallop figures. If one side puts aside ethics and reason for political power, then the other side may choose to do the same. It is the old ethical puzzle: when is one a freedom-fighter and when is one a terrorist? The chance for a reasoned, ethical answer to the abortion question is slipping by. With it also goes the basis for the value of human beings, which has never been politics or law, but ethics. Society needs to begin to think about the philosophic and practical mistake of equating law with ethics, and demanding that objections to law be silenced. The Allies did not want the German people to be silent about National Socialism's lawful brutalities in Germany. We should not expect people to bow before a political power to silence them on the core question of whether a fetus deserves the same ethical respect as you and I. That question has to be asked. Neither current side may like the answer, but the answer is more than process ethics. It must be content ethics.

HUMAN WORTH AND MEMBERSHIP IN THE ETHICAL COMMUNITY OF MAN

The argument made by postmodern bioethics is that we cannot have valid content and must settle for political process. One would think that the abortion debate illustrates this contention best, and that is the assumption that most of my students make when we begin an anlysis of abortion. Most are concerned about polemics, about their position not being accepted, and about being overrun by group dynamics. That does not occur, to their surprise and relief, and for that reason, I remain optimistic about society's capacity to analyze this ethical question and come to a reasonable answer. We start with the basic question: what entities have membership in the ethical community of man? I am willing to start anywhere on the board, and it has to be a large board, because the end result is a long line showing a continuum on which some decisional and definitional points need to be placed. In any human thinking process, we need to start with provisional categories.

Some in the seminars wish to have a human person, a member of the ethical community, begin at conception, one of the provisional categories. When a sperm with half the human complement of DNA successfully penetrates the egg with its half of human DNA and sets in motion the biological process, some wish to acknowledge membership. Even though this is a potential viable entity requiring the initial circumstances of conception and the assumed continued development in an appropriate environment, some feel that potentiality is sufficient and reasonable. The genetic blueprint, although requiring ongoing proteins and specified environments, is their sole requirement for membership. There are, in the course of drawing this continuum, problems on either side of this conception point that make it less appealing, however. Continuing reductively downward and using the same potentiality, initial conditions, and continuing appropriate environment, one could just as easily argue that a sperm cell or an egg cell are also members of the ethical community – and historically, this has been done, with an emphasis on not spilling "seed" or creating barriers for the egg. Granted, these cells are half-complete unless we draw in the potentiality principle, but they will be complete at fusion, and why is this different that the embryo's developmental requirements? Some are now willing to place the point at germ cells. However, germ cells and somatic cells both contain human DNA, and a liberal application of the potentiality principle can move the point to any human DNA (with transforming stem cells into neurons or with cloning techniques using somatic cells, this becomes less theoretical and more scientific). By now, the seminar can see the trend in progression over the continuum. DNA, after all, is DNA, the same four base pairs are shared with all life forms on this planet, plants, insects, fish, birds, mammals, bacteria, with the only exception being viruses with RNA. We can, of course, go to ecosystems with this framework, and the seminar by now has appreciated that ethical issues in "animal rights" and general environmental ethics may hinge on where we reasonably place the point of member. Also, the conceptual problem with potentiality and initial conditions as the single defining concept for membership becomes clear. After all, we could even go to star systems and galaxies, or at least the Gaia hypothesis, with this, and make the human species the single guarantor of ethical obligation for the cosmos, with no reciprocal obligation. That is too great a burden and might truly fit the definition of hubris.

We could limit membership by insisting on human DNA and its conception process, although by now the students would understand that this is a pragmatic limitation that must be justified by showing why it too should not be part of the eliminated potentials. For those who wish a deductive or absolute ethics, the exercise is over. It is on this point that postmodernism has latched, and drawn a conclusion which does not follow, that no content ethics is possible. Rather, there is no need to demand a deductive or absolute ethics when a probable and pragmatically-justified one will do as well or better. By "pragmatic" I do not mean the feel-good definition of postmodernism, or the misinterpretation of situational ethics to mean a rationalization for doing what one wishes. Casuistry was pragmatic and worked better than its opponents give it credit, while still holding to general ethical principles. As in symbolic logic, any formula is certain and precise, logically necessary, until one tries to plug some real-world meaning into the symbols. Then everything breaks loose.

Are there aspects of conception that require it be maintained while germ cells are dropped from membership? We now have to go in the other direction of the continuum. There is an unusual process that may take place from about 14–21 days after conception, and though it may seem trivial at first glance, it has important implications (12). During this time period, if we have decided that the fertilized egg is *a* member, and call him A, we may get ourselves into a conceptual conundrum. Our A may do something unexpected, namely split into two units, twinning, and we now have B and C, or do we have A′ and A″, or do we have A but with two forms? Theologians have worried about when the soul is placed in the material organism with some good reason, because granting membership at conception and having one soul will now be problematic at 16 days when we need two souls (or can we get by with half a soul?). Philosophers concerned with identity also have a conceptual problem, because A = A is no longer true. My suggestion follows the suggestions of others that we wait until that period is over before assigning membership and identity to the changing embryo(s). If we place the point after 21 days, it will make an important conceptual difference, our category will make more sense, and a very important issue in bioethics may be resolvable to most parties arguing it.

Research on human embryos is generally done within that same time period, and IVF is also safely within that time frame. If the pre-21 day embryo is not a member of the ethical community, then such research can be ethically done in that regard, and much of reproductive medicine will no longer be an ethical issue about manipulation and destruction of a human entity to which we owe the same ethical obligations as to an adult member of the community. There will still be issues about respect for human biotic material, and how disrespect could impact on perceptions about human worth. There will still be the reality that we are dealing with a human genetic blueprint, and the need to acknowledge the worth of human genes. Those ethical rules can be in place without requiring a concept that gives membership in the ethical community to an entity whose identity itself is in question. If we insist on having a category that pays no attention to these conceptual problems, then we had best be prepared to extend membership in the ethical community infinitely, because potentiality can be argued step by step across the cosmos. The students in the seminars have found these conceptual discussions useful, expanding, and non-threatening. They have not been polarized into political opponents with the constant threat of coercive activism from both sides. The first steps into a full discussion of abortion can be made. The journey, however, is much longer, and we have merely set the ground rules for free philosophic analysis.

The next points on the continuum involve nonviable fetuses and fetal development. This is the most difficult field of discussion, and I postpone it until we have looked at the further reaches of the continuum. Some bioethicists have argued that infanticide may be ethical, especially in the time period between birth and three months, so membership in the ethical community could be deferred until after three months or after six months (13). One of the reasons for rejecting membership tends to be the assumed lack of cognitive development, and philosophers do tend to want to define membership in terms of reasoning ability, as Kant did or Joseph Fletcher attempted in his definition of a person (14, 15). Of course, such a strict criterion would actually exclude many other entities: developmentally disabled, mildly retarded, comatose and persistent vegetative state, demented, delirious, addicted, school dropouts, mentally psychotic, anyone legally incompetent or lacking capacity to consent. If we take the principle of a rational moral agent, like the potentiality principle, it may extend much further than acceptable, if the only definitional principle.

We will see in Chapter 7 how our culture has actually begun to extend this principle from ethical issues at the beginning of life to ethical issues at the end of life, and eventually anywhere inbetween. The students were reluctant to use the categorizing because of the irrational results of excluding most of humanity from membership at one time or another. The point had to be moved away from infanticide and to the point where a legal person traditionally had been defined, delivery. However, with the inception of partial birth abortion and the recent court ruling that it was a constitutional right, the seminar was now in conflict with the culture. Not only did almost all students wish to hold a firm point at birth/delivery to qualify for membership, they wished to consider that a viable fetus was not conceptually different from a delivered baby. They did not accept the ethical sophistry that, while a live birth was a full member, killing the full-term fetus after only partially delivering it meant there was no live birth, and hence, no member that had been killed without justification. Again, there is a caveat here. We do kill, sometimes ethically, members of the ethical community. What the students objected to was killing the entity halfway through delivery, and then arguing sophistically that a dead entity had been aborted. If the killing is ethical, we can call it what it is and its ethical justification will hold.

The point of membership was now moved toward an earlier time, with agreement that a full-term fetus (unless suffering from a lethal disease incompatible with life outside the uterus, and thus not viable) was a member of the ethical community. Even here, some wished to eliminate that exception and include all full-term fetuses. The next ethical principle was viability, and again there was general agreement that a viable fetus was a member of the ethical community, very much in keeping with the original legal thinking of Roe v. Wade before it was reinterpreted by the court to reflect an absolute right to privacy, even in the third trimester. Viability did not mean capable of full, independent life without support systems of some sort. Intensive Care Nurseries were seen as similar to caring for a full-term baby, with such care required for survival. That the care was highly technological did not make a sufficient difference. The issue with viability as a principle was determining when the fetus was viable, and that is a changing situation. When Roe v. Wade estimated viability, it was 24 weeks. Developments in medicine have moved that timeframe down to 19–20 weeks, when it is possible for a fetus delivered at that time to survive with much technical assistance. Morbidity is always a concern, and there is no guaratee of a healthy baby without residual effects from the prematurity, but viability has extended back in time to approximately 20 weeks, where, pending the research and development of an artificial uterus, it may extend even further back. Viability is also influenced by the presence of lethal genetic diseases such as thanatophoric dwarfism or polycystic kidneys, so development is not the only factor. But diseases may not be a heuristic concept for defining membership in the ethical community. Many adult members of that community do have eventually terminal diseases, and we are all eventually terminal, so it does not seem a workable concept. None of us would be members. Preference should be given to the developmental time and its implications for the possibility of life outside the uterus.

The viability of the fetus was accepted as one of the ethical principles, forming a difficult balance with the potentiality principle. So far in the seminar discussions, political rights or the woman's absolute right to choose had not come into play, because it is first necessary to determine the status of the entity and then to apply other ethical rules which refer to the interactions of two or more members of that community. It is necessary to know what we are talking about to do philosophic ethics. At this point on the continuum, we had conflicting placement of the point that defined membership in the ethical community: 21 days after conception, and 20 weeks or changing timeframe of viability. In fact, the problem area inbetween is the area where social consensus will be difficult and where ethical arguments must be made about the status of the entity that will change the conception in the minds of reasonable human beings. It is also the timeframe in which abortion is used as a replacement for either sexual abstinence or contraception. As such, it is the actual field of controversy. That controversy will grow when the ultrasound three-dimensional, photographic quality pictures of 16 week fetuses become common. Having seen 20 week fetuses in our autopsy room aborted because of lethal genetic diseases, I know such entities, although visually different than a full-term baby, are sufficiently similar to make a prima facie case for membership. We cannot safely overlook that response which recognizes an entity as somehow a member of the species. It is a biological response that is very powerful, as Lewis Thomas argued, very biochemical (16), and we ignore it at our peril as a social species. In what Roe v. Wade called the first trimester, however, the membership is not that apparent and potentiality can only go so far in making an argument. One does not construct an oak chest from an acorn (17). From 8 weeks to 16 weeks, there are insufficient conceptual categories to conclude there is full membership in the ethical community and the balance should lean toward the entity we are reasonably sure is a member, the woman. Her needs and interests should take precedence, but they need to be established, not given as an absolute civil right to privacy. There are ethical and unethical abortions during this time frame, and I see no good argument why the state cannot require sufficient good reason to take the life of any life-form, even one within the uterus. It is not an undue burden to require some reason for killing, if simply to affirm the culture's valuing of all life. The woman's needs and interests would have priority, because it would also be unethical to harm her for no good reason, and the existence of a non-viable fetus does not seem a good reason for harm, if she has real needs and interests. In fact, a short required discussion about her position on the ethics of the abortion might be a preventive to psychological consequences of the abortion afterward, and would achieve something valuable for her. It would be preferable to the political solution of misrepresentation (public relations) and coercion (power). There would still be those who disagreed with abortion during this time period, but their concerns would at least be validated and addressed, if not fully, at least partially. The woman would be educated to the awesomeness of decisions to terminate life and the culture would have reaffirmed that, despite varying positions on ethical membership, it was not a culture of death but a culture of life.

The courts, however, have not made it easy for us. Eventually, the seminar looks at the other member of the ethical community and her obligations to the entity within her as well as its responsibilities to her if it too is a member. The fetus, as a member, would have certain responsibilities. Historically, it was recognized as an aggressor when the mother's life was endangered by it, and it was dealt with as an aggressor would be in self-defense or war. The fetus is not an innocent in consequences, only an innocent in intentions, and that distinction is important. The law has been very contradictory about the legal status of a fetus. In many cases, a viable fetus does have legal status and charges of murder or child abuse may be brought against another person. In abortion, the recent rulings give no legal status even to a full-term fetus.

The law gives only qualified constitutional rights to members of society, except in abortion, where women are given an absolute right to privacy. The seminar discussed the ethical implications of such an absolute right, given in the form of sexual discrimination, and given on a core issue for the very existence of the state: who is a member of the state and who must have defense. In ethics, such an absolute right, even to take away the potential rights of another being, is unprecedented. Even the individual's right to his life is qualified. For me to be given the choice of whom I wish to see as a member of the ethical community or state is a power I should not have. It is an individual ethical relativism beyond what an existentialist would approve, since even that radical ethics qualifies with the concept of authenticity. It is also a dangerous political power, restricted it is true to its application to a fetus, but dangerous because of the precedent set. It is based on defining one group as entitled "victims," and allowing that group because of victimology to victimize another group, a terrible logical inconsistency and an ethical horror story. If such power can be constitutionally given to women, it can be given to other groups, and applied to other entities than fetuses. In fact, it has been, in the business world, in academia, and in community institutions where the life of the entity has not been taken away, but where the free thought, expression, or economic survival of the entity has. It contributes, by corrupting the law, to the creation of the total society envisioned by postmodernism on a global scale.

I recently wrote about my experience with obtaining an abortion when it was illegal, and the letters of response were again interesting, from both positions in the debate. Dr. W. D.Gutowski (18) appreciated my honesty and my admitting that there was a mourning period (and still in some sense is, which is as it should be). He described it as a "separation" or "death of a child." In fact, I agree with him that it was the loss of my potential child, and that the loss occurred because of my decision. I think it was an ethical decision and I did not take it lightly. He wondered if I would have made the same choice if the baby were two years-old, would I have killed my child to save my career path. Notice that he assumed the fetus is a baby, and if that were established, of course my answer would be certainly not. There was more involved than a career path as I hope I fully described in the columns, but even with all those other factors, the answer would still be certainly not. His point is that, since he believes the fetus is a baby, that there can be no good reason to justify killing a child. Dr. Gutowski has got the philosophic question right. It is just that he assumes the answer too easily. Still, I prefer someone who has at least gotten the right question and is discussing ethics to someone who is a political activist ready to dismiss all ethics as irrelevant. As if ethics could be irrelevant to killing!

Dr. Mark Krieger still does that. In his letter based on the same columns, he again proves the postmodernist ideology that has incorporated bioethics (19). First, he engaged in class warfare, a standard political strategy, by implying that persons of intelligence and affluence could easily make the underground connections for an illegal abortion. He is on the side of the common person, the victims of coat hanger abortions, the non-English speaking, the poor women, the uneducated – all the victims selected by this ideology. That I would have qualified as "poor," that my family was the common blue-collar class, that the father and I were truly working our way through state colleges (I as a Five and Dime clerk, butcher's assistant, and finally TV librarian) would upset his preconceptions of who was and who wasn't a victim. Dr. Krieger's point on the continuum for membership in the ethical community isn't until full delivery, and until then, he believes only the woman can decide if this entity lives or dies. The woman is the sole authority, an amazing gift of authoritarianism to women. He then goes on to describe any limitation on this absolute right to be an "injustice" leading us down the old slope where "conservative, male-dominated, political and religious factions have held sway." For good measure, he talks about "right-wing politicians and fundamentalists" who threaten this right of women. He states clearly that this right is an absolute right, the absolutism being required to prevent anyone from imposing their concept of ethics on the woman. Then he produces a typical postmodern argument that the fundamental problem is there are no absolute ethical standards, with the implication that ethics is simply varying invalid notions and thus of no importance in the abortion question or any other social question. Notice how political the tenor of this letter was, and how specifically ideological the politics are. Those are the unverified, invalid assumptions of postmodern bioethics as applied in the real world and as contained in the theoretical discussions of postmodernism.

WHEN WOMEN MAKE THE IDEOLOGICALLY-INCORRECT CHOICE

In 1997 our hospital had its second case of a complicated pregnancy with persistent vegetative state or brain dead pregnant women. The first case was in 1996 and involved the rape of a young woman in a nursing home who was in a persistent vegetative state. The woman's mother had to decide, when her daughter's pregnancy was finally diagnosed at 19 weeks, whether to continue it or have an abortion. The male aide with previous charges of sexual abuse and impersonating a sex therapist was found guilty based on DNA evidence. The mother made her decision after seeing an ultrasound of the fetus and partially based on her and her daughter's religious beliefs. She decided to have the physicians do everything to continue the pregnancy. A premature baby boy with some health problems was delivered. Less than a year later, as I walked through the airport to a meeting of the American Academy of Psychiatry and Law where we were doing a presentation on this case, I saw a news headline that the young woman had quietly died at her new nursing home.

The second case was in 1997 and involved an unmarried 20 year-old who was in her second trimester of pregnancy when she suffered a brain aneurysm. She met the legal criteria for brain death, and the physicians had her on life-support while they discussed with her mother what to do. Her mother was at her bedside as she became emaciated on life-support, and her mother's first instinct was to remove the support. The father did not want two deaths and preferred that support be continued until the fetus had a chance to survive. A court hearing was scheduled to give the mother legal decision-making authority, as the fetus reached 23 weeks. The mother also saw an ultrasound, and like the mother of the first case, cancelled the court hearing and decided to maintain her daughter on life-support as long as feasible. The realistic limits to such support were no more than 10 weeks until death occurred even on support. The fetus, if all went well, would make it to 26 or 27 weeks. The young woman actually made it further than physicians had predicted, and gave birth to a 9 week premature baby girl who was healthy and beautiful as all babies are. What was the bioethics response to this case, which like the other, received national attention from the media?

I could not have made up better evidence for the politcizing of bioethics within a postmodern model. The response was very critical of the two mothers' decisions to keep their daughters alive and continue the pregnancy. In the case of the brain-dead woman, the director of a bioethics center told the media that this was research using a human subject who could not give consent, unethical and "mad-science experimentation" on a corpse. Informed consent, impossible for a brain-dead woman but possible for her surrogate, her mother, was seen as more important than giving the fetus some extra weeks of development without any harm to the pregnant woman. Another bioethicist who commented raised the spectre of using/abusing the woman as an "incubator," and saw the case in terms of the woman being denied her free choice about the pregnancy, her privacy right. For some reason, this bioethicist thought the woman would have chosen to abort (the young woman did not seek an abortion before the aneurysm in her second trimester, in reality) and that she should be taken off life-support. But it was the bioethicist who pigeon-holed the woman into a pro-choice for abortion position. The local media gave only one view, the ideologically-correct one, and selected out the two bioethicists who found the choice to continue the pregnancy to be unethical. The same objections and almost the same phrases were used for the first case. The concern was that the young woman was not being given her choice to abort, although every indication was that her belief system would have favored continuing the pregnancy and that that would have been her choice. For postmodern bioethics, both mothers made the incorrect choice for their daughters. The implication was that abortion was preferable and should have been chosen. But the surrogate decision makers did not choose abortion, and that somehow was seen as an attack on the young women, taking away their rights and reducing them to objects, incubators.

One could argue that these are statements for the popular culture, not what bioethics does in its academic journals and texts. One would be mistaken. Also, bioethics has always been an activist, media-involved field, as Wikler has admitted (20). The media statements accurately present postmodern bioethics, and what they present is a political ideology and a political feminist agenda within that ideology. The statements are not valid ethically, but accepted politically.

First, keeping a brain-dead patient on life-support to give a fetus the chance at viability is not experimental or research. Medicine already knew what it needed to apply this technology and knew its limitations. Medicine has good clinical experience with comatose, persistent vegetative state, and brain-dead patients, some of whom were pregnant. There was no mad-scientist experiment going on, no Dr. Frankenstein, no crossing of new borders, just fairly routine medical care that was sensationalized by the media. Furthermore, the brain-dead woman was not declared dead, therefore not a corpse, not some accumulation of dead parts lying on the research table in an old castle. She had a name. Her mother chose to keep essential functions going because her daughter's uterus was still functioning to sustain the life of her fetus. She was still a person, not an object, unless we chose to make her one. With her past history, she still deserved human respect, most of all from the bioethicists who should not have demeaned and degraded her. Politically, however, the individual is a unit that can be used for the cause, with few restraints on that.

The argument that this woman was an incubator was the bioethicist's perception, not the mother's. The mother watched her dying daughter, worried over her, discussed what the best thing to do for her would be. She also saw her daughter as a mother-to-be, with a potential that was now a viable fetus, carrying her daughter's genes, part of her daughter that would go on into another generation. It is only when ideology forces the perception of exploitation of women that such a wonder cannot be appreciated or considered. Pregnancy itself is devalued and carrying on genetic information is given no value at all. Ideology can't and won't allow for that. Its focus on political power, on dominance and submission, on oppressor and oppressed can't see or deal with such biological values. It is singularly unable to fully understand femaleness. It is that reality that makes any choice but a choice for abortion the ideologically-wrong choice. There were good ethical reasons to support the two mothers in their decisions to continue pregnancies, but the ideology would not allow it. It is not an ideology of free decision making. It also cannot ground human worth, except for the instrumental use of human beings toward a social cause. For those with no such utility, the outlook is grim.

THE DIFFICULT QUESTION OF PERSONAL IDENTITY AND MEMBERSHIP IN THE ETHICAL COMMUNITY

No matter how our ethical discussion attempts to avoid describing the nature of the entity or entities in question, natural experience has a way of putting the question to us over and over again. In September of 2000, the British Royal Courts of Justice had to rule on just such a case. The three-judge court turned down an appeal of a lower ruling by the parents of conjoined baby girls, and allowed physicians to proceed with a separation attempt that will sacrifice one of the girls in order to save the other (21). The girls were given pseudonyms to preserve confidentiality, Jodie and Mary. They were born August 8, 2000 in Manchester where their Roman Catholic parents from Malta had come to get the best care for the impending conjoint twin birth. Mary has no heart or lungs, and depends on Jodie for circulation. Her brain is undeveloped, and she is joined to Jodie from the pelvis, with only one pair of legs. There is an 80–90% chance, the treatment team says, that both Mary and Jodie will die in the next three to six months, because Jodie's heart cannot support two growing bodies. Although stable now, Jodie is failing to grow under the present stress. There is also no hope that Mary could survive if separated from her twin, since she relies on her for heart and lung functioning.

The parents cannot accept that one of their children should die to allow the other to survive, nor do they feel they could agree to killing Mary. Their ethical position is to depend on God's will and accept even the likely death of both children. However, the state is taking away their parental authority to make this decision. Their parental love and duty, and their theological beliefs about the relationship of humans to the cosmos, are being set aside. The Archbishop of Westminster, the Most Rev. Cormac Murphy-O'Connor, submitted a written statement to the court, describing Mary as "innocent" and enjoying the same "basic worth and dignity" of every human being in the eyes of the Church. He said she had done nothing which could justify killing her, and separation was therefore not morally permissible. There was an assumption that Mary had actually not done anything to justify killing her, and that could be argued. The basic assumption, however, is that Mary is a human being with full standing in the ethical community.

The judges struggled with this question. Could Mary, incompletely-formed and depending for viability completely on Jodie, be a person who was born alive, "a person in being" entitled to protection from the law in order to live out her natural life, no matter how short that might be. One of the lawyers compared Mary to a person being kept alive artificially on machines, and argued that physicians would not hesitate to feel it was ethical to remove the patient from life-support in that circumstance. One of the judges decided that Mary had a right to life, but then with words took that right away, saying she had little "right to be alive" since she was alive only because she was parasitic on Jodie. Jodie was the one whose right to life needed to be protected from her sister who was killing her. If this is a straightforward "self-defense" argument, giving both girls the right to life but taking it away from the aggressor, it is an argument that works. Mary does not have the right to kill Jodie, and the only way to prevent it is to separate the twins, even if the unwished for consequence is Mary's death. Here, both girls are treated as full members of the ethical community, Mary's unintended aggression is also recognized, and intervention is proposed that as one of the effects, will be the death of Mary. If other means could be found to protect Jodie's life, they would be used. Since there are none, unless one is a complete pacifist and rejects killing a human absolutely, the operation would be ethical.

But there were undertones of a less humanistic base for deciding. Jodie's law advocate did dehumanize Mary by claiming that no purpose would be served in keeping unviable Mary alive: ". . . there are no interests in preserving what is unfortunately a futile life." Here we see the concept of futility in its current communitarian sense. The value of life depends on serving some state or group purpose, and if the entity cannot serve an acceptable purpose, it finds itself without standing, without value. We can then, it is concluded but a bit hastily, do what we wish with that entity.

Still, this does disturb some sense within us, and it is easier to make this argument if the entity, Mary, is described in less positive ways. For that reason, the argument starts with "unviable" and goes on to "assault", "sucking the lifeblood of Jodie," "parasitic living, "beyond help," and "undeveloped brain." One of the specialists, contradicting the early death prognosis, thought it was possible the twins could survive longer, perhaps even for years, with Mary slowly weakening Jodie until both died, picturing Mary as a terrible mistake of nature that remained attached to the normal child while slowly draining away her life. Mary is now of less value than Jodie, who has more standing in the ethical community. It is the felt need to dehumanize Mary that presents the conceptual problem. Her parents do not dehumanize her. Their theological perspective treats her with respect. The judges do not really wish to dehumanize her. Still, to grant physicians the legal right to proceed with the separation and kill Mary raised an awful specter in their minds. Lord Justice Ward asked: "Do we save Jodie by murdering Mary?"

At this point, the case becomes much more problematic. The prognosis does not seem quite so scientifically-verified, although I suspect it is actually accurate. But much prognosis is not accurate, and despite this fatal flaw (literally), many medical decisions of life and death are made on such tenuous prognoses. The dehumanizing of Mary is unnecessary and implies that bioethics does not support the innate value of a human being, that value has to be earned and can also become too expensive to maintain. The judges' decision does not seem to face the central question: When is it ethical to take a human life? The state's intervention in the parents' decision is troubling. The parents are not described as legally incompetent in any way, or as malevolent. They can articulate why they do not wish the operation to occur. These are their children, and an argument for child abuse has not been made, nor do I think could it be made, even if Jodie will die as a result of the decision.

The decision of the judges reflected the unfinished nature of the ethical analysis. Lord Justice Ward felt Mary was living on borrowed time, incapable of independent existence, and "designated for death." He saw Mary as a dying member of the ethical community, and was willing to hasten/cause her death in order to save Jodie. He did not feel she had a right to be alive. Using that same argument, unfortunately, neither does Jodie have a right to be alive. She is attached by nature to her twin and will soon be incapable of existence herself, designated for death just as her twin is. The opposite argument could be made from Justice Ward's decision, that there should be no intervention for either child. Both children could be beyond ethical help.

Lord Justice Brooke had a photograph of the two girls and wondered how any separation could be done without its being a criminal assault. He countered that argument by suggesting Mary was criminally assaulting Jodie, and that Jodie could withdraw consent for her circulatory system to be used for Mary. Nature, however, did not ask Jodie for consent, and the doctrine of informed consent seems stretched here. Justice Brooke finally decided the separation would give the children's bodies the integrity which "nature denied them," overlooking the fact that it was the court that would be denying the natural integrity of conjoint twins and that nature did not seem too precise about integral identity. The argument seemed to be that we should accept Mary's death in order to give Mary bodily integrity, but that seems an irrational argument.

Even membership in the ethical community will not avoid these bioethical issues, but it can help us avoid rationalizing. The Mary/Jodie case does not represent well-done ethical analysis, but rather the increasing acceptance of the state's intervention in parental decisions and a nascent utilitarianism that requires all of us to demonstrate our social utility. It is sometimes ethical to kill a member of the ethical community, and if we faced that reality honestly, our arguments would be constructed more validly. Although it is sometimes ethical, it is never routine or easily accepted, and the arguments required for ethical justification must be very strong arguments.

THE OTHER END OF THE ETHICAL COMMUNITY CONTINUUM

We have been talking about ethical questions revolving around the beginning of life, and where on the continuum to put the earliest point for membership in the ethical community. That continuum, however, is longer than the point of birth and delivery, as the two cases of the severely brain-damaged pregnant women illustrate. Are they still members of the ethical community? Can one lose membership because of certain events or does a history of membership imply continuing membership? Is a brain-dead woman a member still? What is the ethical status of a corpse? Is it ethical to kill an adult whose membership in the ethical community is granted? Is there again a privacy right to make any medical decisions at the end of life that the patient wishes, and is there an obligation for physicians to facilitate those decisions, not only through physician-assisted suicide, but even with direct termination or unnecessary terminal sedation?

The opposite end of the continuum in the abortion ethical analysis turns out to be the analysis of ending life at its approaching biological end, and the same political strategies and ideological assumptions hold for this bioethical issue as well. Again, I will argue that postmodern bioethics has politicized dying as well as reproduction.

REFERENCES

1. Krieger, M. (1999). Abortion is about rights, not ethics. *The Medical Post*, *35*(40), 12.
2. Wilson, M. (1999). Abortion is about a woman's right to control her body. *The Medical Post*, *2*(23/1999), 17.
3. Millar, T. (2000). Abortion discussion "ignored" ethics debate. *The Medical Post*, *1(9/2000), 13.*
4. *Watson, T. (1999). MDs have somehow, somewhere lost their moral way. The Medical Post*, *11*(9/1999), 17.
5. Hume, D. A. (1888). *A Treatise of Human Nature: Being an Attempt to Introduce the Experimental Method of Reasoning into Moral Subjects*, L. A. Selby-Bigge, (Ed.). Oxford: Clarendon Press.
6. Fletcher, J. (1966). *Situation Ethics: The New Morality.* Philadelphia: Westminster Press.

7. Taylor, R. (1973). *Action and Purpose*. Atlantic Highland, NJ: Humanities Press.
8. Stenberg v. Carhart. (June 28, 2000). U.S. Supreme Ct.
9. Greenhouse, L. (2000) *Court rules that governments can't outlaw type of abortion*. New York Times.com/library/politics.
10. Planned Parenthood of S. E. Pennsylvania v. Casey. 112 S. Ct. 2791 (1992).
11. Hilts, P. (6/30/2000). *Doctors express relief over decision: Abortion foes vow to continue fight*. New York Times.com/library/national.
12. Marquis, D. (1999). Why most abortions are wrong. In: R. B. Edwards & E. E. Bittar (Eds), *Advances in Bioethics* (pp. 215–244). Stamford, CT: JAI Press.
13. Tooley, M. (1983). *Abortion and Infanticide*. New York: Oxford University Press.
14. Kant, I. (1952). *The Foundations of the Metaphysics of Morals* (Translated by Beck, L. W.). Indianapolis: Bobbs-Merrill.
15. Fletcher, J. (1974). Four indicators of humanhood – The enquiry matures. *Hastings Center Report*, *4*, 4–7.
16. Thomas, L. (1995). The Lives of a Cell: Notes of a Biology Watcher. New York: Penguin: and presentation at University of Rochester School of Medicine.
17. Clements, C. D.(1982). *Medical Genetics Casebook: A Clinical Introduction to Medical Ethics Systems Theory*. Clifton, NJ: Humana Press.
18. Gutowski, W. D. (2000). Is it ever right to end life of a child? *The Medical Post*, 2(22/2000), 15.
19. Krieger, M. (2000). The wealthy have always had access to safe abortions. *The Medical Post*, 2(22/2000), 15.
20. Wikler, D. (1994). Bioethics in health policy: What methodology? In: E. Bittar & N. Bittar (Eds), *Principles of Medical Biology*, Vol. 1A, pp. 23–36. Stamford, CT: JAI Press.
21. Hoge, W. (9/23/2000). *British court orders surgery to separate twins*. New York Times.com/2000/09/23/world.

7. POLITICIZING BRAIN DEATH, TREATMENT REFUSAL, PHYSICIAN-ASSISTED SUICIDE, TERMINAL SEDATION, TREATMENT RATIONING

CREATING THE CULTURE OF DEATH

One of the first articles of mine that was published was an analysis of the ethics of suicide that, from the perspective of the person considering the act, argued that the choice for non-being, non-existence, obviated any possibility of ethics for that person (1). My argument was that the person considering suicide had rejected the basis of ethics (human needs, interests, purposes and meaning), and that ethical questions no longer applied to that person. I think now that it was selected by the editors, one an activist for euthanasia in all the forms allowed by The Netherlands, because it appeared to give broad philosophic justication for suicide. My understanding of suicide has increased greatly since that writing, and I can see how it would be very easy to misinterpret my too-benign description of the choice for suicide. It was really a plea for understanding on a philosophic level the position of a person who no longer wished to exist in this world, and whose wishes reflected a philosophic understanding of the human condition. What it overlooked was the difference between struggle and withdrawal in the human condition, and accepted the suicidal person's resignation to sad realities as the only option. In other words, the article itself began with the acceptance of the suicidal person's response to reality and with the renunciation of ethics because of that response. It might have been a partial understanding of the suicidal individual's internal dynamics, but it was much too relative to that individual and the personal conclusions reached. It was only partial in describing the internal dynamics, as Freud underscored in his concepts of thanatos and eros. The internal struggle between life-affirming and life-denying that the article did discuss did not give sufficient weight to that struggle and the objective analysis of both options. It did reflect the probable truth that the cosmos is neutral to either the yea-saying or nay-saying option, to use Nietzsche's phrase, and that dynamic systems are always in the process of either coming together into greater organization or breaking down into smaller components. Yet it missed the possibility for humans of making evaluations about life-affirming or life-denying choices beyond individual or cultural relativism.

In its attempt to bring understanding to the suicidal act, it missed the wonder of the human condition in favor of the despair of the human condition. Sometimes, suicide can reflect both, but rarely, and these are generally not the suicides that bioethics discusses. There is a marble statue in the Museo delle Terme in Rome of a Celtic warrior in the act of suicide. It is a truly magnificent piece that catches the human condition in all its grandeur. The statue is a copy of one of the bronzes on a monument commemorating the battle of Pergamum, where the Greeks defeated the Celts/Gauls, and is a companion piece to the more famous statue, "The Dying Gaul." This statue is of a warrior surrounded by his Greek enemies, about to be captured and likely sold into slavery. His wife has fought with him, and he has just killed her to protect her from that fate. With one strong hand he holds her slumped figure so that she does not fall to the ground, protective to the last. With the other, he holds his sword to his clavicle, his head turned to the advancing Greeks in a last expression of human defiance and freedom. The sword has penetrated toward his heart, with marble drops of blood beginning to flow. The statue contains love, pride, defiance, resolution, strength, courage, dignity, and human worth, all caught in a piece of marble. Beside that, in a strange way, life-affirming act of suicide, my early argument for the passive, renouncing, withdrawing act of life-denying seems unworthy of human history. It too readily accepted the suicidal person's reasoning for thanatos and emotional response toward death.

Yet it was no surprise that it was published, because postmodern bioethics promotes a choice for thanatos, an emotional response moving toward death. Whether it reflects or has itself introduced a death wish into the culture, a nay-saying that turns everying toward nihilism, is not something that can be answered here. What is important is that it is creating a culture of death in a very fundamental sense, very much like the suicidal person my article tried to understand and give worth to. Even its political ideology is designed to annihilate a people and their history, on the general social level, and a person and his or her history, on the individual level (2). Perhaps the best way to understand this culture of death is to look at how bioethics has changed our society's approaches to dying and death.

CERTIFYING BRAIN DEATH

One of the first changes made by postmodern bioethics was the addition of a new criterion for death, brain death.Technology had partly caused this development through the routinized use of scarce organs for transplant, and through life-support systems which could for a limited time keep the body's organs from failure even though the natural mechanisms for such functioning were no longer working. It would be a mistake to completely credit medical technology for this ethical issue, however. Even before the 1960s and 1970s, a dying body could be partly sustained, and there had always been a concern that the criteria for death be precise enough not to mistake rare events for death (non-detectable pulse and respiration). The need for organs was real, and the capacity to keep those organs oxygenated was innovative. The best chance to get organs was from an otherwise undamaged body which had massive brain trauma. Nevertheless, I will argue that there is much more to the campaign for a criterion of brain death than need for organs, that organ donation was the good end to justify a more far-reaching purpose.

In the late 1960s, the Dean of Harvard Medical School asked its Committee on Human Studies to redefine death and that became known as the Harvard Brain Death Committee (3). Its report was published in the *Journal of the American Medical Association* and became the model for many state laws redefining criteria for death. The Committee looked at ways of determining when the human brain had been so completely damaged that it ceased to perform any of its functions. There are a number of techniques to make that determination early, but some are clouded by support technologies. If there is true brain death, the brain stem cannot, for example, regulate temperature or control respiration and death is inevitable but not necessarily speedy if support systems are used. There are tests to determine if there is blood circulation in parts of the brain and they are reasonably reliable. An electroencephalogram can be done at three different times over a period of days to determine if there is any sign of activity in any part of the brain. The legal definition of brain death is whole brain death, and even with life-support systems, there is a limited amount of time before complete organ systems breakdown and the most obvious signs of death, decomposition. "Whole brain death" is an important concept and was specified initially to make certain that the actual process of dying had neared completion, so that brain death would be the equivalent of no respiration or cardiac function. The potential for abuse of this definition was recognized in the legal statutes, but laws are primarily defined by the regulations that flow from them, and by the actual practice in the real situation.

The term "brain dead" is now often used inappropriately for persons who have severe damage to the frontal or mid-brain areas, who may be in a coma or even a persistent vegetative state. Unfortunately, clinicians are not as precise as the legal statutes, and families may be told their family member is brain dead when that term does not meet the legal and medical strict definition of brain death. It may seem a small medical step from whole brain death to partial brain death except for the brain stem, but it is actually a huge conceptual leap with serious implications. Although organ donations would not be procured from a person who does not meet the standard of brain death, the families may make decisions about removal from ventilators or continuing aggressive treatment based on such a looser definition. In the case of anencephalic babies who are dying but not legally yet brain dead since there is some inadequate functioning of the brain stem, the wish of the parents to donate a heart or another organ, and the wish of clinicians to also give some meaning to the anencephalic baby's brief life by saving another child may result in a brief cessation of life-support. Brain death can then be certified by a pediatric neurologist, and then the rapid reinstatement of support to keep the organ viable is done. Loma Linda Hospital has tried such a technique to procure heart transplants for infants born with hypoplastic hearts, and a Canadian couple donated their anencephalic baby Gabrielle's heart to another couple's infant, Paul, so that this infant could survive. At no point in this process was Gabrielle seen as an "it." She was a baby who died and gave the gift of life to another child. The hospital did not feel it was necessary to deprive an anencephalic infant of membership in the ethical community, but that raised problems about the ethics of turning off life-support for a dying baby in order to make a subsequent determination of brain death.

There was criticism in the bioethics community about these transplants because the anencephalic infants did not really meet the criteria of brain death (e.g. Alexander Capron), and the procedure is no longer done. That criticism, however, has been muted and has disappeared from recent cases. For example, the mother of an anencephalic infant had requested that full medical support be given to the baby until it died. The argument there was that medical futility precluded the mother's surrogate decision making for her child, and that life-support could be removed. Complicating the issue is that there are degrees of neural tube defects, ranging from anencephalics non-viable in the uterus to encephaloceles which require medical treatment to live (4), and the distinguishing line is not always crystal clear. Nevertheless, bioethicists made the futility argument for removal of life-support. In addition, the strategy of redefining anencephalic infants as non-members of the ethical community, similar to a strategy to redefine persons in persistent vegetative state as non-members, is a utilitarian extension of the concept of brain death to now include partial brain death. Such persistent vegetative state patients, such as Nancy Cruzan, have been referred to in briefs to the U.S. Supreme Court as "it" (5). If the entity, whether Gabrielle or Nancy Cruzan (although those doing the redefining prefer not to use personal names), can be redefined as a non-member, then the same ethical obligations we have toward others who are members do not apply to these entities. There is then no valid reason why the medical system need care for them or refrain from terminating them. This is a very different approach than the previous arguments about organ donation from anencephalic infants. Those arguments depended on the infant being declared legally dead for the purpose of then taking the organ. There was no need seen to redefine the infant as a non-member outside the human community, an it. But the new arguments, weak if only justified by medical futility, do require redefinition. The goal is to remove medical care from the infant with or without the parents' permission, but without making the medical system appear to be in the business of terminating patients without their or their surrogate's consent. The only connection between the organ donation goal and the cost-containment/efficiency goal is that both deal with a dying entity but not with the question of how to define that entity as a member or non-member of the ethical community.

Still, the trend to expand the narrow definition of brain death is one that needs to be watched. In practice, it is already occurring, even if in legal statute the narrow definition prevails. Bioethicists are arguing for an end to the requirement of whole-brain death, the next incremental step (6, 7, 8). More disturbing for its demonstration of loss of human feelings is the recent debate in Great Britain, where anaesthesiologists have finally spoken up against the practice of removing donor organs from the brain-dead person without the use of anaesthesia. The experiences of watching blood pressure and heart rate shoot up as the scalpel cut, of seeing the body writhing and moving so much that the surgeons then ask that the person be paralyzed because they cannot operate, have traumatized nurses and physicians. However, it was only after guidelines were published by the Intensive Care Society saying anaesthesia was not required for brain-dead persons that the journal of the Royal College of Anesthetists wrote publicly, breaking the silence previously maintained so as not to discourage organ donations (9).

The re-interpreting of "brain death" is not occurring for the practice of organ donation directly. The impetus driving the trend appears to be cost-efficiency in a downsized health care system, and Chapter 8 will discuss the economic aspects of postmodern bioethics in detail. For now, the important point is that if the entity can be redefined as dead because of partial brain death, the assumption can be made that a corpse is not part of the ethical community (although I will argue that this assumption is false). If an object is not part of the ethical community, the conclusion can be that society can ethically approve of saving money by disposing of such entities, even though the surrogates object, using the concept of futility. This ethical argument might not be required under triage conditions in an undeveloped country, where human beings who are firmly recognized as members of the ethical community must die because of real limits to resources. In a developed country, that argument cannot be successfully made. The current shortages in access to care, facilities' resources, staffing shortages, and research and development do not reflect a natural shortage such as the Club of Rome postulated (10), but an artificial shortage created by political choices about allocation and prioritization, and postmodern value choices that devalue human beings who do not have special victim-status.

Unlike the Gabrielle case, which had a positive outcome for her parents and the parents of the baby boy, a gift of meaning and commemoration for the tragedy of the anencephalic child, the recent medical futility application of bioethics is a terrible harm to the family. The parents have already suffered as a result of the neural tube defect and the wish to deny that their baby is dying. Should their desperation be allowed the expense of time to resolve into facing reality? The bioethics' futility argument does not even consider that ethical question in its disguised cost-efficiency standard. The medical system compounds the grief of the situation by becoming adversarial, fighting with the parents about terminating their child, not for the best interest of the anencephalic baby, but for cost containment. For all the discussion in the beginning of bioethics about humanizing medicine, what now becomes apparent is the cold, economic, institutional neglect of human nature. A full cost/benefit analysis is not even attempted, because the bioethics' conclusion has already been assumed. However, the long-term harm to parents and other surrogates of such a confrontation with the health care system cannot be set aside in the equation. It is a legitimate cost, every bit as much as the daily cost for the hospital room. The cost of the change in the ethics of the society also cannot be set aside in the equation, and it is likely it will create a world that is neither liveable nor sustainable, no matter how much wealth is saved for reallocation to other social purposes, some good, some banal. In systems methodology, as we will see, the unnecessary sacrifice of sufficient numbers at any level will destabilize the system at all levels. Since our world of experience is largely composed of dynamic natural systems, it would be wise to pay attention to systems analysis.

The concern with the brain death criterion in our current postmodern culture has been that it will, by barely perceptible incremental steps, with sufficient probability for serious ethical concern, expand in practice to include any brain trauma that renders the human being who was once a member of the ethical community as now a non-member.

The ethical continuum of membership can again be useful to ask the needed ethical questions. If an entity is once recognized as a member of the ethical community, can it be removed from membership? In other words, is there a historicity principle which requires recognition of continuing membership and its obligations, even when that entity can no longer perform its ethical obligations and has little social utility? This still does not logically imply that we must never kill such entities. It means that we must have strong reasons for such a killing, and that we must acknowledge that it is a killing of a member of our ethical community.

Surprisingly, when the seminars look at the end-of-life on the continuum, students have the same problem placing category-points as they did for the beginning of life. It at first seemed simple to place the point before a death certificate or a dead body. A corpse, it at first seemed, could not be a member, because it was dead, a thing. The same assumption generates the trend to redefine some patients as no longer persons, but things, its. That strategy may not be as successful as its proponents would like. In analyzing the status of a corpse, two examples tell a different story. One is a current one from the "peace action" in Somalia. The other is an ancient Greek tale about Antigone.

When U.S. marines attempted to enforce the national policy of preferring one Somali war lord over another, they attacked the southern war lord in the capital of Mogadishu. Using Black Hawk helicopters, the marines quickly met more resistance than anticipated, and a helicopter went down (11). Attempts to rescue the men resulted in even more problems. Some marines on the helicopter had been killed, and the angry residents of that section of Mogadishu dragged one dead marine's body through the streets. Television crews captured the partly-blackened dead body being pulled through the dirt of the city. The typical response was horror at the abuse of the body. Why, however, did we use ethical terms like "abuse" and have valuational feelings like horror and disgust, if a corpse is assumed not to be a member of the ethical community? The dead marine felt nothing. Why did almost anyone watching that news clip feel an evil thing had occurred? Why do men at war put their own lives at high risk to bring in their dead from the battlefield? My ethical argument would be that past membership in the ethical community is also present membership based on a memory principle, that human minds remember each other, hold each other from the past to the future, even when the body system has broken down in death. Cultures across time have respected the dead, have deemed desecrating corpses to be an assault and an ethically-questionable act.

Turning to Antigone, we see the same ethical issue of respect for the dead body (12). In the Greek city of Thebes, Antigone's brothers had led a rebellion against the tyrant, and like so many battles for freedom, it had been lost. Her brothers were killed, and their bodies lay on the field of battle. The tyrant, to make an example and solidify his rule, ordered that no bodies should be buried and that anyone who attempted to bury the bodies would themselves be buried alive. Greek culture placed a large ethical obligation on proper respect and burial of the body. There is a classic painting showing the wife of a Greek general who had angered his ruler, been killed, and his body left to rot in the open. His wife and her maidservant have come to the body, and his wife will eat his body so that he is safe within her and will not suffer the horror of decomposing on the ground. We may shudder at her resolution to this ethical problem, especially after recalling the Fores in Chapter 5, but it illustrates the depth of the ethical rule and the wife's love. Antigone is also up to the task. She does bury the bodies of her brothers, despite the penalty. She forfeits her life, being killed by burial alive, in order to do the ethical thing. Were all these people, from the Greeks to the marines of Mogadishu, incredibly foolish about ethics? I would argue they weren't, that in fact they had a deeper understanding than postmodern bioethics about the good of man and his or her place in the natural world of which we are all part. Classic humanism had concluded that the historicity and memory principles were needed ethical concepts in determining membership in the ethical community, and that was likely based on the sensient nature of man. Given that we hold loved ones in our minds, it follows that membership in the ethical community is a permanent status. For justification of that premise, all we need do is look at the energy spent across cultures and times in trying to hold the dead with us: the kurgans of the Altai, the tombs of the Valley of the Kings, the Incas' ancestor mummies, the Romans' masks or effigies of their families' dead on parade, all the cemeteries and ashes. I once interviewed a Russian emigre physician who chose the U.S. town he established his practice in because of its well-cared-for cemetery. He told me that a town that cared for its cemetery proved its social caring, proved it was a decent place to live. We might want to consider what a loss of the respect for a corpse, a removal of membership, might indicate about the ethics of a culture that has moved in that direction.

THE NEXT INCREMENTAL STEP: REFUSAL OF MEDICAL TREATMENT

Rothman documents that in 1974, an article by Dr. Walter Modell appeared in the *New England Journal of Medicine*, " On Medical Intervention" (13). This was the introduction of a bioethics concept about "advanced directives," which generated full support within the bioethics consensus. The need of advanced directives still had to be shown to the public in a way that would garner their support. In 1976 this strategy became active with the trial balloon case of Karen Ann Quinlan soon followed by editorials in professional journals, other cases, Sissela Bok's *New England Journal of Medicine* opinion piece, "Personal directions for care at the end of life" (14), and state legal statutes of Living Wills and Do Not Resuscitate Orders. A book also appeared, *The Long Dying of Baby Andrew* (15), which was a case for removing life-support treatment or not instituting it in the first place, and which had the support of bioethicists Peter Singer and Helga Kuhse (16).

The best known of the trial balloon cases is the Quinlan case in New Jersey. The intimate legal connection of bioethics with this case is illustrated by the work of Robert Veatch, then of the Hasting Institute (a bioethics think tank) and later director of the Kennedy Institute (the second major bioethics think tank). Veatch helped prepare the lawyer for the Quinlan family for the court. The New Jersey Supreme Court ruled that the Quinlans could decide to remove their daughter Karen from her ventilator. Karen was not comatose, but in a persistent vegetative state, and there were many indications she did not actually need a ventilator for respiratory support. In fact, when she was removed from the ventilator, she breathed adequately on her own and eventually died in a nursing home after many years. Her parents did not petition the courts to remove hydration and nutrition. The court ruling, however, allowed patient or surrogate to request life-saving treatment be removed or not instituted. Hospitals were encouraged to set up committees to review termination decisions, generally staffed with hospital administrators, hospital counsel, chaplains, bioethicists, physicians and other medical staff.

Massachusetts courts then allowed treatment refusal in the Saikewicz case, and New York courts in the Brother Fox case. The *New England Journal of Medicine* published Charles Fried's editorial, "Terminating life support: Out of the closet" (17). The ethical argument made by bioethicists was that a competent patient could either directly refuse medical treatment or request it be removed, or could write an advanced directive that needed to be honored by physicians and hospitals. Not providing medical treatment or removing it was only to be done at the choice of the competent patient, or in accordance with the advanced directive. The family could not override this request; the medical system could not override this request.

In the following years, the standard for competence dropped to a narrow legal standard that generally requires orientation to time, place and person, and minimal understanding of the consequences. Although state laws often required an evaluation for the presence of psychiatric illness such as depression, I have served on ethics committees sufficiently to know that a sensitive evaluation is rarely done; that patients involuntarily committed to psychiatric facilities because of a serious illness that makes them a danger to themselves and others are accorded competence to make such choices; that dementing patients can be assumed to be momentarily competent to make such decisions; that serious depression is missed; and that small, ambivalent gestures are frequently interpreted to mean a wish to discontinue treatment, even life-saving treatment. Compounding this problem, patients are asked over and over again to produce a living will or a DNR order, and U.S. law now requires that on admission to hospital, patients receive a living will form and are encouraged to fill it out. My hospital experience also indicates that patients are structured to choose for treatment refusal or withdrawal. Three actual examples of structuring are:

(1) the old oil rig worker with lung cancer who had a nurse visit him every day to ask if he wished to have a DNR order in his chart. He kept saying no; the nurse kept asking;
(2) the standard question to an older patient about whether the patient wishes to suffer broken ribs and burned skin by being resuscitated, or wishes to simply be allowed to die;
(3) the young Afro-American woman shot and left quadraplegic after a drive-by shooting, who was told with the ventilator she would never be able to talk again, and that there were no rehabilitation facilities in the area for her to go to, but that if she wished to have the ventilator removed, it could be.

Treatment refusal rights were quickly extended to patients who had not made out a living will, but who might have said casually to a friend or family member that they would not wish to be kept like a "vegetable." Family could now make decisions based on their perception of what the patient might have wished. Parents could make such decisions as surrogates for their children.

Although laws often explicitly stated that DNR orders and advanced directives were not to compromise comfort care or care required for pain control, in practice such effects are frequent. Admission to an intensive care unit is less likely if either DNR or advanced directives are in the chart. Surgery to relieve pain, as in the case of a young girl with Hurler's syndrome which is eventually terminal, was problematic for two reasons: the mother had placed a DNR and advanced directive in her chart, and the condition would make anaesthesia higher risk. The patient went out-of-state to get the pain-relieving surgery, but the mother had to be extremely assertive to find a hospital which would do it.

Patients whose families wish treatment continued and who have not left a directive requesting treatment be withdrawn or not instituted are now being classified as medically futile cases and hospitals are going to court to have treatment withdrawn or refused. The Baby L case is one example where the mother wished treatment to continue, the baby was not terminal in any immediate sense, and the hospital refused based on medical futility (18). Both bioethicists and the courts supported the hospital's position, although the mother was very devoted to this baby and could give home care most of the time. It was only in crises that the child required hospitalization. Another hospital offered to accept the baby, or this would have been a case where a parent requesting her non-terminal child receive life-saving treatment would have been refused such treatment and would have had to watch her child die unnecessarily. Postmodern bioethics here claimed both medical futility and child abuse, since the treatment could be uncomfortable for a brief period of time and since the child could not achieve a partial cure or acceptable (to the bioethicists) baseline from treatment. The legal argument was that the courts could not force professionals to make particular decisions, such as whether to treat or not treat, that this was a question of expertise, one of the few times physicians are granted such control in bioethics. That this rare time involves cost-efficiency issues is instructive.

The Wanglie case is another example of a woman who had requested all medical care, lapsed into a coma, and then had her surrogate family's decision to continue care fought in the courts by the hospital, based on medical futility and economic concerns.

The history of The Netherlands' experiment with euthanasia needs to be carefully written by a good historian. It is the history of an incremental development, from competent patient to incompetent patient, from painful and terminal illness to any illness (even anorexia), from physical illnesses to organic brain diseases and finally to psychiatric diseases like severe depression, from passive euthanasia to active euthanasia, and now the growing fear of the elderly that it will be without consent (19, 20).

In the U.S., I know of an AIDS case where the HIV-positive physician wanted everything done medically that could be, and where his medical care team wished to place a DNR order. He refused to have a DNR. There were economic and psychological pressures in this case, and the team met with the parents to discuss a "silent DNR" (the young man would not be told he was DNR). The hospital ethics committee considered the economic issues relevant and was willing to entertain the option of a silent DNR. Perhaps fortunately for bioethics, the young man, not originally seriously ill with his first opportunistic infection, slipped into a coma, and he was allowed to die before forcing the hand of postmodern bioethics. Here the advanced directive was clear. The patient, as well-informed as any patient could ever be, had asked not to have a DNR, and had requested that all medical treatment be given. The advanced directive was not honored by the ethics committee, because like continuing a pregnancy, it was not the ideologically-approved advanced directive.

It is not that hard to make sense of this contradiction. I was given the opportunity to do that many years ago. The hospital ethics committee was meeting to look over the state regulations that interpreted the state's new DNR law, and to see which of multiple forms it would make sense to use. There was a rainbow of colored forms. The committee had an interesting discussion about the state's obligation to protect its citizen's lives when there was a compelling interest in doing so. The new law indicated that the state did not have a compelling interest in patients in hospitals making decisions that would end their lives. The citizen's liberty interests overruled the state's protective obligation – in the hospital. I walked to my car in the parking garage with the hospital counsel. We were having an animated discussion, and caught up in the interesting ideas of the moment, I got in my car and forgot to put on my seatbelt, probably the second time I had ever done that. I drove two blocks down the city streets and saw a traffic check ahead of me. Knowing my tires were nearly new, my papers were in order, nothing was wrong with my car, I waited my turn. The officer looked things over and then suddenly noticed I did not have on my seat belt. He began to write me out a ticket. I protested that I thought I had it on, when he asked me why I hadn't simply clicked on the belt before he got to the car. It did me no good, because the state believed it did have a compelling interest in keeping me from making life-threatening decisions – on its streets and roads. Driving home, I tried to reconcile the two events. The contradiction disappeared when I considered that an economic purpose underlay both situations. It was cost-efficient for terminal patients or chronically-ill patients to die sooner rather than later, so refusal of life-sustaining medical treatment was economically beneficial. Medical costs begin to rapidly accelerate within the last three months of life, so eliminating that expensive period of time is quite cost-efficient. On the other hand, medical costs of car accidents could be lowered if driver and passengers wore seat belts, and most drivers still have some social utility. Preventing serious injuries through seat belt use is also cost-efficient. Moral autonomy, ethical choice or civil rights have little to do with any of this. In the next chapter, I will make the argument that economic first-order decisions have little to do with ethics and values, and that much of ethics and values, as Calabresi and Bobbit argue (21), is the strategy of disguising these hard, tragic economic decisions so that they do not offend our stated cultural values.

In 1990, the U.S. Supreme Court refused to interpret an absolute right to privacy in end of life decisions, and further ruled that in order to remove hydration and nutrition, now considered as properly belonging in the category of treatment refusal, there must be some reasonable evidence that the patient would have chosen to refuse food and water. In the Nancy Cruzan case, another trial balloon, the bioethics trend was to support such a privacy right in decisions between patient and physician and to allow surrogates to make the decision to remove hydration and nutrition. It was in this case that a bioethicist/neurologist referred to Nancy Cruzan as an "it" (22). Nancy Cruzan was in a persistent vegetative state, a new diagnosis that is quite vague science and that extends the concept of brain trauma from comatose individuals to those who have sleeping and waking cycles, sometimes can take food orally but are usually tube-fed for convenience, may have pain responses, and some vocalizations. Some track movements with their eyes. The level of awareness is in serious question, however. There is no doubt that serious brain damage has occurred, and it is rare that some PVS cases regain sensience, although some do. The individual with PVS is not terminal, with nursing care. Is this individual a member of the ethical community? If on the basis of historicity and memory, ethics can ascribe membership as ongoing, then yes, Nancy Cruzan was always a member of the ethical community and in the memory of those who knew her, still is. Is it ethical to terminate her existence if she left no evidence that is what she would wish? In this case, the Supreme Court gave her legal status and required some evidence that she had discussed what she would wish if she were PVS, since that was felt to be a legal (and ethical) obligation that the community had toward her. New evidence was then somehow found from friends who stated that she had discussed briefly her repugnance at imagining herself kept alive in such a state. There was no written documentation; the evidence was not really clear and convincing, especially coming so late in the trial and only after the court had demanded it; and the interpretation of what she had casually said is open to question. However, the state of Missouri found, after having fought the case through the courts to continue nursing care, that there was now sufficient evidence that removing food and water was what Nancy Cruzan would have chosen. Hydration and nutrition were removed, and she died some days after. It was now bioethically acceptable to withdraw hydration and nutrition for an incompetent patient, with no documentation required.

KEVORKIAN, QUILL, AND PHYSICIAN-ASSISTED SUICIDE

The incremental progress of euthanasia and the right to die did not stop with treatment refusal or withdrawal, which is still regarded as primarily passive euthanasia or indirect killing. The first trial balloon case was Janet Adkins, a woman who was in the early stages of Alzheimer's disease and whose family supported her wish to die (23). Since medical treatment was not required for her to live, treatment refusal or withdrawal or withholding was not a means to her death. Dr. Jack Kevorkian, a pathologist who had left many positions and was no longer practicing, but who held a medical license in Michigan, had begun to propose that patients be allowed to have help from physicians in committing suicide. Mrs. Adkins heard about Dr. Kevorkian, contacted him, and eventually became his first assisted-suicide. At this time, Dr. Kevorkian was attempting to stay within the legal framework that would not allow direct killing, so he had constructed an apparatus that the patient could trigger and that would deliver lethal drugs. He and Mrs. Adkins, after a brief meeting and discussion, and after he had reviewed her medical records, went into a van where he had set up his apparatus. Mrs. Adkins, with her own hand and with Dr. Kevorkian's hands not touching the apparatus, then committed suicide. The death was reported and investigated, but the Michigan authorities were unable to make a case against Dr. Kevorkian.

This first assisted-suicide set in motion an accumulating number of deaths reaching into the forties. Patients with multiple sclerosis were over-represented, but the diagnoses included individuals with reported uncontrolled pain, cancer patients, patients with amyotrophic lateral sclerosis, and one patient with no diagnosis but a history of addiction and psychiatric treatment. The latter patient was very problematic for ethics, but apparently not for postmodern bioethics. She was the ex-patient of a physician who had fallen in love with her and married her. There were undocumented reports of marital problems and parenting problems, as well as addiction problems. The domestic problems could have proved damaging to the physician husband, because even though he had married her, such relationships are forbidden in the new postmodern bioethics. Even as his wife, she could have brought malpractice charges. Dr. Kevorkian accepted her for his assisted-suicide method, which now had changed to the use of carbon monoxide since his medical license had been taken by the state of Michigan and he could not write prescriptions.

Still, society could not seem to deal with Dr. Kevorkian, and criminal charges were never successful in preventing his continuation of the assisted-suicides. Physicians had become uncomfortable with Dr. Kevorkian as a model for the compassionate provider of help with suicide, and with his clinical expertise. The criticisms were that he had never had a clinical practice and did not have an adequate physician-patient relationship with the people who came to him for help. Bioethicists who supported assisted-suicide, and the consensus was in support of it, were uneasy, not with the ethics, but with the frightening image that Dr. Kevorkian presented for the public. With his support eroding, Dr. Kevorkian made a radical move. He agreed to be interviewed by Mike Wallace for television, and presented Wallace and eventually the public with a taped death that was not assisted-suicide but direct physician termination of a person. It had always been Dr. Kevorkian's argument that assisted-suicide was a temporary step to the goal of direct termination on demand, not necessarily by physicians but by specialized professionals in clinics. He felt there was a constitutional right to die and that the individual could demand facilities and direct killing within them. He argued that refusal to supply the means for such deaths was unethical. Deserted by bioethics and public relations, and having clearly violated Michigan law on the tape, Dr. Kevorkian was this time convicted of manslaughter and sentenced to prison.

The bioethics consensus had already floated another trial balloon, the "Debbie" case in the *Journal of the American Medical Association* (24). This was an anonymous piece, and perhaps a made-up case, since its facts could not be determined. A young resident supposedly was called to the room of a nearly-terminal young woman dying of cancer and in pain. There was an older woman in the room who was never identified in the piece. The resident gave the chart a quick read, having never before seen this patient. In the conversation with the young woman, she reportedly told the resident "let's get this over with," and he inferred she meant she wished to die quickly. It is not clear from the piece what this phrase really meant as it is open to many interpretations. The resident thought, however, that it was clear. He went to the nursing station, did not have a consultation with the physician on the case or any other professional, prepared an injection which would depress the patient's respiration and lead to painless death, and went back into the room. He did not discuss with the young woman or the other person in the room what he was intending, but gave a direct injection which then caused death.

This case was not well-received. The Illinois district attorney wished to know the name of the writer, and the editor refused to give the name. Bioethics discussion did not develop into any workable agreement. However, it represented a case-pleading in a leading medical journal for consideration that direct injection by a physician which caused the death of a patient might be ethical. That in itself was a notable development in the history of postmodern bioethics, and it shone a light on the final goal of euthanasia advocates within bioethics.

The bioethics consensus finally centered on another, more public-relations friendly, trial balloon. This case appeared in the *New England Journal of Medicine*, written by Dr. Timothy Quill, the "Diane" case (25). It was followed a year later by another article in the same journal suggesting guidelines for physician-assisted suicide (26). "Diane" was a woman in her late thirties who had been seeing Dr. Quill about eight years. She had had a history of alcoholism and depression, but was doing very well in the biopsychosocial treatment of behavioral medicine. She was bright, articulate, becoming self-fulfilling in her own business, a "good" patient. When she developed a rash, the blood work indicated leukemia and she was devastated. Since the disease would require specialists and hospitalizations, the physician-patient relationship she shared with Dr. Quill was now at risk of change. For her reasons, Diane refused treatment for the leukemia, and since Dr. Quill was once director of the hospice program, she was comfortable continuing in his care. Although a bone marrow transplant was one intervention, Diane did not want the technological intervention. Nor did she want the pain of allowing the disease to kill her. She told Dr. Quill she would want to kill herself when the pain or other issues became too great. This opened a new working relationship with Dr. Quill. She did rounds with him on her case, co-teaching with him in a sense, a colleague supporting his beliefs on medical overtreatment. He also referred her to the local Hemlock Society, reinforcing her decision.

When Diane called Dr. Quill requesting a prescription for barbiturates to sleep, she lied in order to protect him from legal consequences. However, he told her he knew the real reason she wanted the pills. He met with her and told her the lethal dose, asking her only to inform him before she killed herself. He wanted to see her for an appointment before the act. She did call him, they met and talked about meeting again in another life at her favorite location, a lake, with dragons swimming in the sunset. Two days later, she used the pills he had prescribed for her, after sending her husband and son out, and died. The family called Dr. Quill, he called the medical examiner's office and reported the death of a hospice patient from leukemia, and there the case would have ended, except for the journal article. Interestingly, in that article he used the phrase about dragons swimming in the sunset twice.

Because "Diane's" name was not her real one, the matter would have gone no further. Bioethics now had its acceptable trial balloon. The physician was a respected clinician in the community, and had a compassionate, empathic air about him. He had known his patient for years. The woman was terminal because she had refused medical intervention that was her only chance for survival, but the refusal seemed well-reasoned. Time had been taken with the case. It was discussed in teaching rounds. Dr. Quill's article was well-received and he was promoted as the ethical alternative to Dr. Kevorkian.

Even after the anonymous tip that led to the identification of "Diane" and the discovery of her body which had been donated for nursing student education, the legal process favored Dr. Quill. The coroner did ascertain that the cause of death was overdose of barbituates, and Dr. Quill had illegally lied to the medical examiner about cause of death, but no action was taken on that. A grand jury was convened, but the assistant district attorney did not recommend a grand jury finding and the only witness was Dr. Quill. The grand jury did not choose to indict, although the case was a violation of New York State law. Following resolution of the legal problem, the activism for assisted-suicide increased. Dr. Quill was co-author of another *New England Journal of Medicine* article that proposed guidelines (27). The title of that article did not refer to terminal patients, but to "hopelessly ill" patients, and a terminal disease, as in treatment refusal or withdrawal, is not required. The article talks about "incurable conditions" of "unrelenting suffering." Again, incurable conditions are not necessarily terminal conditions and could be developmental defects or disabilities. "Suffering" is not physical pain. It could be unhappiness, depression, the expression of understanding of the human condition. A great deal is made of the difference between assisted-suicide and direct killing, but here we should be careful. If a physician patterns the patient's response to disease by actively supporting or encouraging suicide, if the physician supplies the means for the patient, if the physician gives the message of approval through frequent meetings and talk, if the physician wants to be notified and present when the act is done – then the physician and the patient are both carrying out the suicide. The physician is directly responsible for the death, even if the patient's hands are the final instrument. That would be the philosophic ethics analysis. The law would take the opposite position and make a legal distinction.

The guidelines dismiss any possibility of coercion of the patient, any economic force that would create pressure for the use of assisted-suicide, and instead focus on the harm of prohibiting assisted-suicide such as an undignified death and mental suffering. The guidelines also do not rule out the future acceptance of active termination of patients. They call for monitoring changes in public policy which might allow such physician terminations of patients. Leaving that option open is telling. Physician-assisted suicide will not be the final goal of bioethics.

Guidelines are meant to safeguard against abuse, but there is nothing in these guidelines that would actually work to protect individuals from such abuse.

Even the specific clinical criteria show how little the patient would be protected. One criterion is that the patient must have an incurable condition associated with unrelenting suffering, but we have seen that this could include almost any condition and need not involve any physical pain, only unhappiness. The criterion specifically states the condition need not be terminal. The second criterion is that physicians should ensure that suffering and the request for suicide are not the result of inadequate medical care. Physicians cannot ensure this in current health care systems, and those systems will continue to make it probable that there will be inadequate medical care. This guideline is wishful thinking in the real world of economic pressure. A third criterion is that there should be clear and repeatable requests. This criterion has already failed for treatment refusal or withdrawal. There is no reason to think it would work for physician-assisted suicide. The guideline itself states that the patient should not have to beg, so it is internally inconsistent. A fourth guideline is meant to insure that the patient's judgment not be distorted by emotions, and that a psychiatric evaluation may be given if the primary physician feels one is needed. From a practical perspective, there are not enough psychiatrists to do such consults, nor are the expenses likely to be sustained. It is furthermore no safeguard at all, since the decision is in the hands of the primary physician about calling for a consult. A fifth guideline suggests the assisted-suicide be carried out only in the context of a meaningful doctor–patient relationship. That will be hard to do in a care delivery system where continuity of care is slowly being eliminated and there is no physician power to change that system. A sixth guideline says another physician should check off on the legal competence and autonomy of the patient, on diagnosis and prognosis, and on informing about alternatives to assisted-suicide. This paperwork will not safeguard from abuses since it will devolve into a routine check-off for the charts. A seventh guideline requires clear documentation, again not a safeguard against abuse. An eighth guideline recommends informing the family, but makes the point that this is not required. This isolates the patient by removing the natural advocates for the patient. The family, even when notified, has no legal power to prevent the assisted-suicide even if they strongly disagree, providing no safeguard. Finally, the rationale for the guidelines is that there is less risk if we routinize assisted-suicide than if we have the current condition of rare personal and professional decisions between doctor and patient. That conclusion does not logically follow. Making something routine, legal, and bureaucratic does not lead to safety for anyone. It generally leads to abuse.

If the delivery system were understood, these guidelines would be seen as ineffectual, as previous guideline safeguards have been. Postmodern bioethics was not concerned with the futility of safety guidelines, however. The support for physician-assisted suicide became the bioethics consensus, a final solution for the total society. Cases began to move through the legal system in the U.S., and eventually from the U.S. 2nd and 9th Circuit Courts of Appeals (28) to the U.S. Supreme Court. Majorities in both circuit courts sided with the plaintiffs who wished to establish a constitutional right to die, but who also wanted state laws against assisted-suicide struck down. Judge Guido Calabresi voted with the majority in the *Quill, Klagsbrun, and Grossman vs. Vacco, Pataki, and Morgenthau* case on April 2, 1996. Justice Calabresi was concerned about making any all-embracing judicial decision, one way or another, and told our forensic group he hoped the U.S. Supreme Court would make a very limited decision on the issue. Like many judges, he was reluctant for the courts alone to make this decision and felt society might not yet be ready for a universal answer. He agreed with the majority that the New York State laws against assisted-suicide could not stand, but was uneasy with a due process or equal protection argument to establish a constitutional right to assisted-suicide. He preferred to let state legislatures decide. Of course, the New York State Legislature had already decided, first by its refusal to remove the entire prohibition against suicide by keeping in the language against assisted-suicide, and second by accepting the New York Task Force on Life and Law's recommendation to criminalize assisted-suicide. Calabresi disagreed and listed the escalating laws that New York had passed on treatment refusal, medication refusal, DNR, and the Health Care Agents and Proxies Act to prove there was an acceptance of the basis for physician-assisted suicide. This is a straightforward presentation of the incremental strategy in euthanasia. Calabresi did not see any difference in these various end-of-life options and thought this step-by-step pattern toward euthanasia and physician-assisted suicide (PAS) was now firmly written into law. He saw no reason to draw a line before PAS and felt there was a clear due process violation for patients who wished for active assistance and patients who wished passive assistance.

Most major medical professional organizations had filed briefs opposing PAS, but the circuit courts weren't interested in learning the realities of the medical care system. They would allow, without concern for economic pressures, the possibility of terminating patients who did not have a terminal disease and would not oppose making decisions about termination that were based on economic goals. As an economic utilitarian in legal theory, Calabresi was willing to consider economic goals as part of the decision making process, provided we could present the means to those goals in ways that would not call into question our public values. Bioethics supplies that seemingly ethical language to lull us about the implications for actual human dignity, worth, and freedom in PAS. For example, in Calabresi's concurring opinion, he talked about this issue as "a deeply personal decision," described such assisted death as "quiet, proud death, bodily integrity intact," and the choice for such death as "choices about death touch the core of liberty . . . the essence of our humanity." In his argument before the Supreme Court, Lawrence Tribe used such words frequently, even talking about the human cycle and its key phases.

These terms do not help much. Why must dying be quiet in order to be proud? What in the real world is a physically intact body and what has that got to do with integrity in the human sense? What could the judges have meant in talking about choosing how to die being a deep and personal experience? All our human existence/being is a deep and personal experience if we pay attention to it. What has that to do with asking a physician to help in a suicide? How is the core of liberty and humanity touched by choosing to commit suicide, anymore than it would be touched by understanding that chosen relationships with others require an obligation not to leave a suicide legacy? What does the demand for liberty to commit suicide have to do with demanding a physician must help do it? Does political liberty really mean being able to do anything one wishes to do?

I have no doubt the justices felt the incremental changes they were effecting were progressive rather than regressive. However, I would argue they chose not to hold the line against that most regressive of ideas in bioethics, PAS. Reading the opinions is frustrating, because the real risks within the health care delivery system are made irrelevant or denied. The Quill vs. Vacca case continued to the Supreme Court, where a line was finally drawn at the beginning of 1999 (29). The Court refused to create a privacy right to die. It argued that state laws against assisted-suicide were constitutional and would stand. It left open the door for states to chose, however, to pass laws that permitted assisted-suicide. One state has done so.

THE NEW BIOETHICS STRATEGY: TERMINAL SEDATION

Dr. Quill and other proponents of PAS and direct euthanasia, still supported by the bioethics consensus, have moved to a new political strategy to achieve the long-term goal of unlimited euthanasia. The newest phrase is now "terminal sedation." Terminal sedation is the continuing use of medication that renders the patient completely sedated and is given in either toxic or lethal amounts. There is objective ethical justification for controlling pain in a dying patient, even when control can only be achieved through lethal amounts of medication. In fact, it is an ethical obligation of medicine to provide relief from pain when no other interventions are possible. Abandoning the patient to die in pain is not an option. Achieving pain control is problematic, however, because of regulations that many states have against prescriptions of controlled drugs, and the bureaucratic monitoring of physician prescriptions for those drugs. The war on drugs does have a casualty, the patient who requires pain medication in amounts that are outliers.

One would think, then, that a bioethics position that is activist for terminal sedation would be philosophically justified. Again, there is a caution here. We need to do a full analysis of what the goal actually is. Where in pain control, the goal is to give enough medication to achieve a pain-free state for the patient, and that goal determines the amount of medication used, that is not likely to be the case for terminal sedation. Freud's physician, for example, used first a safe amount of medication to attempt to control Freud's severe pain as he lay dying of oral cancer (30). After a few hours trial, when the pain could not be controlled, the physician increased the dose to a lethal dose. If, however, a non-lethal dose had been able to achieve pain control, as in the first attempt, that would have been the ethical dose. To deliberately increase the dose to a lethal dose when there would be no reason to require it for pain control is simply a carefully-disguised direct termination of a patient by the physician. So we need to be very clear about the difference between justifying lethal doses for pain control when it is established such doses are needed, and going directly to lethal doses for the real purpose of killing the patient. The ethical justification of one does not cover an ethical justification for the other.

This distinction, given the manipulation of the public and the political strategies of postmodern bioethics, is not likely to be maintained. Otherwise, in fact, a new name would not be needed. Why call "pain control even with lethal doses" by a new name, terminal sedation? No, the new name suggests the real meaning and goal. Terminal sedation is just what it says, killing the patient with a lethal sedative. It could be used, for those patients who requested it, where pain control lethal doses could not be used, in non-terminal conditions and for unhappiness/suffering rather than for physical pain control. This terminal sedation would be direct killing through medications. The physician would be putting the patient to death. It would not be a "natural" death where "nature was allowed to take its course," to use the bioethics phrases that attempted to justify treatment withholding, refusal or withdrawal. Those phrases, having achieved their ideological goal of incremental advances toward direct euthanasia, have now easily been replaced in a reversal of the bioethics argument. Now bioethics wants to give patients control, the interventionist capacity to shorten their lives, the help of physicians to achieve a medical death rather than a natural death. The key phrases here will be "control," "dignity," "choosing the time and manner," and "the right to die." The right to die is a political right similar to the pro-choice absolute right to privacy. In fact, the constitutional arguments are the same for both, based on the *Griswold* decision of a privacy right. It illustrates, as do the many lower court and high court decisions about euthanasia, the basically political and ideological consensus of bioethics.

The concept of "control" and "choice" also go back to the discussion about the contradictory stances on autonomy picked by bioethics to suit its ideological premises. The patient is given an illusion of control in the health care delivery system by being granted the control of termination. The patient is given the choice to end treatment and existence.

Dignity is also redefined within this ideology. It is a surface dignity that does not express the inner dignity that is the true source of a human's dignity. A human being in diapers can have inner dignity that is more important and stronger than the superficial beauty implied in present definitions. Inner, reciprocated love can exist and grow stronger in the presence of vomit, urine, feces, tubing, technologies, and delirium. Human life has always been messy, and there have always been those who wished to clean it up, but it is in the messiness that we may find our most fundamental humanity. The cultural opportunities for this actual human dignity and its reciprocal relationships grow less, and as they grow less, the possibility of ethics grows less with them.

We need to take care, then, of the real meaning of terminal sedation, and why it was called up in addition to ethical pain control, which may require the obligation to give lethal doses of medication. That it is being proposed by ideological bioethics, that it is a fall-back position for activists in euthanasia, that it is too much a resemblance to a public relations campaign to change the words or slogans to gain public acceptance, should give us pause.

There is another aspect of this campaign that is troubling. After all, it is not that hard to commit suicide, unless one is a quadraplegic, and even there, one can refuse nutrition. Why should assisting killing the patient or directly killing the patient have even entered the field of bioethics, and from there, law and politics? One can commit suicide painlessly, on one's own. Why involve a health care professional? The argument that such attempts may be botched does not seem impressive. Those attempts that are not ambivalent are generally not botched, as suicide research has shown (31). We have had patients successfully commit suicide even in hospitals. In the world outside, a car and a hose are usually sufficient if medications are not available, and this is not painful or messy. It does have to be done alone, which may bother some, although if the concern is that high, then the argument in that early article of mine applies: the person is not sufficiently committed to a life-denying position to warrant a suicide attempt. If one is not worried about a mess, and there is question whether one should be, then other options work well too. I do not wish to be hard, but I do wish to question the justification for involving the medical care system in patient termination. The risks in terms of the redefinition of that system and the redefinition of physicians are enormous. In Chapter 2, we discussed the change that Hippocratic ethics had brought to healing, the ability to trust the physician, unlike the shaman, not to kill the patient. The loss of that trust would have implications for the healing profession that are not acceptable in any objective ethical system. It is one of the reasons that I argue that postmodern bioethics is very regressive, not progressive at all. In the case of end of life issues in bioethics, it returns us to the Black Shamans. I will say more about that in Chapter 9, but such loss of trust is not to be taken lightly.

The patients requesting euthanasia are often portrayed as sensitive, brave persons who wish to maintain their dignity, who wish to have a loving farewell with family and friends, who courageously play classic music as they die, who value personal control. We should consider that there are alternative descriptions. There is a book on modern gypsies in Europe which tells the story of Jan Yoors at one point, a Belgian teen who left home to join a travelling group of Lovaro Gypsies and lived with them for many years. The book is called *Bury Me Standing* (32). Yoors recalled something his adoptive gypsy father, Pulika, once told him, and we might listen well too: "Too often the courage about dying is cowardice about living."

THE HIDDEN AGENDA: TREATMENT RATIONING

As a preface to the next chapter, I want to suggest what I think is the real politicoeconomic basis for postmodern bioethics' advocacy of full euthanasia, passive and active. As the incident with the seat belt hinted, it would be wise if we took the advice of Watergate Committee member Senator Sam Irwin: "follow the money."

Bioethics justifies, or attempts to justify, the redefinition of quality care into cost-efficiency care, and the utilitarian sacrifice of individual patients for broader social goals. Because its ideology conceives of those social goals, goals requiring a total society to achieve and operate, as more valuable than anything else, it finds disutility in certain individuals and does not tolerate disutility. The incremental move from autonomous choice of passive euthanasia to surrogate choice for incompetents to medical futility and euthanasia without choice is predicated on redistributing resources from the medical care system to other social purposes. It is a powerful political and economic force, a trend that is now institutionalized within the nascent total society and requires no conspiracy theory to activate it. It probably happened with the best of ideological intentions, and the politicizing of most of our social institutions since the 1970s made it easy to recast bioethics in terms of those activist political assumptions.

The group first targeted (in fact, I described this group as having a bullseye painted on its back) was the eventually terminal patient who uses up a third of Medicare resources in the last few months of life. The ideology of bioethics can see no social purpose for these individuals since its concepts are too barren concerning the human condition. Elderly, soon to be terminal patients were the first group to attempt to restrict from access to the delivery system's resources. That move began early, with the director of the Hastings Institute, Daniel Callahan, identifying those individuals who chronologically were no longer members in full standing of the ethical community (33). DNR orders were designed to eliminate these individuals from access to medical care as speedily as could be accepted by the public. Limiting access to expensive medical interventions was also first applied to this group. Advanced directives were promoted for these individuals, as was a natural death without the frightening aspects of expensive medical technologies. When insufficient people signed such living wills or advanced directives, the strategy changed to medical futility. Terminal patients in general joined this targeted group. The group has grown quite large by now, including low-birth weight premies in neonatal intensive care nurseries, those with dementias, brain-damaged patients, AIDS patients, people with chronic diseases and disabilities, and any person whose quality of life (QoL) does not come up to the expectations of postmodern bioethics, expectations which are much higher than those of the patient's family or the patient him or herself. Advanced directives are no longer sufficient to achieve the utility goal that since the 1980s has been a significant part of bioethics, but that was implicitly present from the beginning. In the next chapter, we will see what all of this has done to the medical care system, and the significance of the name change from medical care system to health care system.

REFERENCES

1. Clements, C. D. (1980). The ethics of not-being: Individual options for suicide. In: D. Mayo & M. Battin (Eds), *Suicide: The Ethical Issues* (pp. 104–114). St. Martin's Press.
2. Connor, S. (1989). *Postmodernist Culture: An Introduction to Theories of the Contemporary*. Cambridge, England: Blackwell.
3. Rothman, D. J. (1991). *Strangers at the Bedside: A History of How Law and Bioethics Transformed Medical Decision Making*. New York: Basic Books.

4. Clements, C. D. (1982). *Medical Genetics Casebook: A Clinical Introduction to Medical Ethics Systems Theory.*
5. Cruzan v. Director, Missouri Dept. of Health. (1990). U.S. Lexis 3301 (U.S. June 25, 1990).
6. Truog, R. D. (1997). Is it time to abandon brain death? *Hastings Center Report*, *27*, 29–37.
7. Veatch. R. M. (1993). The impending collapse of the whole-brain definition of death. *Hastings Center Report*, *23*, 18–24.
8. Zaner, R. M. (Ed.) (1988). *Death Beyond Whole-Brain Criteria*. The Netherlands: Kluwer Academic, Dordrecht.
9. Birchard, K. (2000). Do organ donors feel the scalpel? *The Medical Post*, *36*(30), 70.
10. Meadows, D. H., Meadows, D. L., Randers, J., & Behrens, W. W. (1972). *The Limits to Growth*. New York: American Library.
11. Bowden, M. (1999). *Black Hawk Down: A Story of Modern War*. New York: Atlantic Monthly Press.
12. Kelly, E. (1980). Introduction: The ethics of utilitarianism: John Stuart Mill. In: L. E. Navia & E. Kelly (Eds), *Ethics and the Search for Values*. Buffalo, NY: Prometheus Books.
13. Modell, W. (1974). A "Will" to live. *New England Journal of Medicine*, *290*, 907–908.
14. Bok, S. (1976). Personal directions for care at the end of life. *New England Journal of Medicine*, *295*, 367–369.
15. Stinson, R., & Stinson, P. (1983). *The Long Dying of Baby Andrew*. Boston.
16. Kuhse, J. (1999). Euthanasia. In: R. B. Edwards & E. E. Bittar (Eds), *Advances in Bioethics* (pp. 333–344). Stamford, CT: JAI Press.
17. Rothman, D. J. (1991). *Strangers at the Bedside: A History of How Law and Bioethics Transformed Medical Decision Making*. New York: Basic Books.
18. Paris, J. J., Cronte, R. K., & Reardon, F. (1990). Physicians' refusal of requested treatment: The case of Baby L. *New England Journal of Medicine*, *322*(14), 1012–1014.
19. Keown, J. (1995). Euthanasia in the Netherlands: Sliding dow the slippery slope? In: J. Keown (Ed.), *Euthanasia Examined: Ethical, Clinical and Legal Perspectives* (pp. 261–296). Cambridge, England: Cambridge University Press.
20. Van der Maas, P. J., Vander Wal, G., Haverkate, I., deGraaff, C. L. M., Kester, J. G. C., Onwuteaka-Philipsen, B. E., Van der Heide, A. Bosma, J. M., & Willims, D. L. (1996). Euthanasia, physician-assisted suicide, and other medical practices involving the end of life in the Netherlands, 1990–1995. *New England Journal of Medicine*, *335*, 1699–1705.
21. Calabresi, G., & Bobbit, P. (1978). *Tragic Choices: The Conflicts Society Confronts in the Allocation of Tragically Scarce Resources*. New York: Norton.
22. Cruzan v. Director, Missouri Dept. of Health. (1990). U.S. Lexis 3301 (U.S. June 25, 1990).
23. Frontline: The Health Quarterly Special Edition. Choosing Death. Robert Wood Johnson Foundation: PB5.
24. Anonymous (19). It's all right, Debbie. *JAMA*.
25. Quill, T. E. (1991). Death and dignity: A case of individualized decision making. *New England Journal of Medicine*, *324*(10), 691–694.
26. Quill, T. E., Cassell, C. K., & Meier, D. E. (1992). Care of the hopelessly ill – Proposed clinical criteria for physician-assisted suicide. *NEJM 327*(19): 1380.
27. Quill, T. E., Cassell, C. K., & Meier, D. E. (1992). Care of the hopelessly ill – Proposed clinical criteria for physician-assisted suicide. *NEJM 327*(19): 1380.
28. Quill, Klagsbrun, & Grossman v. Vacco, Pataki, & Morgenthau. U.S. Court of Appeals for the Second Circuit. (1996). U.S. App. Lexis 6215.
29. Washington v. Glucksberg, and Vacco v. Quill. (1997). 117 S Ct. 2258.
30. Shur, M. (1972). *Freud: Living and Dying*. New York: International University Press.

31. Clements, C. D., Sider, R. C., & Perlmutter, R. (1983). Suicide: Bad act or good intervention. *Suicide and Life-Threatening Behavior, 13*, 28–41.
32. Fonseca, I. (1995). *Bury Me Standing: The Gypsies and Their Journey.* New York: Vintage Books.
33. Callahan, D. (1989). *Setting Limits: Medical Goals in an Aging Society.* New York: Simon and Schuster.

8. HEALTH BELIEF MODEL AND HEALTH DELIVERY SYSTEMS

THE BEGINNING OF TAKING CONTROL OF THE PURSE STRINGS

When I began in bioethics I confronted two trends. The first was Van Potter's work on bioethics which attempted an actual interdiscipline between biology and ethics and which was instrumental in my entering this field (1). The second consisted of seminars at the medical school in the department of preventive and community medicine, which looked at insurance reform and delivery system changes. One of these seminars was co-taught by the CEO's of the major insurance systems in our community, and led by Dr. Ernest Saward who had been brought from the Kaiser-Permanente health maintenance organization (HMO) to legitimate and set up such a managed care system in our area. Rochester's connection with political policy on medicine had its roots in Marion Folsom's work with the federal government. During the seminar, a prospective payment plan was described and advocated as a necessary reform for the waste in medicine. The plan worked by appealing, as the seminar admitted, to the belief in physician greed. At the beginning of each fiscal year, "x" amount of dollars would be allocated to cover medical treatment of the patient population in the plan. If physicians could lower costs, the extra dollars would be divided among them, at least temporarily. If physicians ran over the amount allocated, the dollars would have to come from their earnings. However, after two years, the "x" amount would be reviewed, and if physicians had consistently come in under that amount, the "x" amount for the following year would be lowered to reflect the actual cost. Now physicians would be faced with the same two alternatives. They could continue to cut costs and still enjoy dividing the extra dollars at the end of the year, or they could run over budget and pay out of their earnings, which were becoming suspiciously like salary. The cycle would repeat itself in two years, with the "x" dollars reflecting the lowered expenses of the previous year, and the physicians again faced with the economic (and ethical) decisions about lowering medical dollars spent even further than the years before. This payment scheme troubled me. I raised an objection in the seminar that it seemed to me a downward spiral in medical care was being created, and I failed to see where the brakes on this road down the mountain were.

Here the two opposing assumptions became quite clear. I thought it likely that costs would be cut through downsizing facilities, staff, access, diagnostic testing, time with patients, treatments offered, hospitalizations and lengths-of-stay, and research and development/technologies. That was my best-case scenario. My worst-case scenario also seemed to me more likely. I thought this downsizing would not be enough, and that certain very expensive groups of patients would be targeted for treatment refusal, treatment withholding, and euthanasia. Physicians would be structured into economic gate-keeping and treatment-rationing. Dr. Saward did not think this would be the case, because, he assured me, there was so much "fat," so much inappropriate medical treatment, so much waste, so much unscientific choice for procedures, that no brakes were required. That did not reassure me. At some point, I thought we needed brakes, but for the life of me (literally, and for others as well) I could not see how prospective payment schemes had any safeguards of that sort built in.

We have come a long way since that seminar, and it has all been a little-controlled careening down that mountain. Dr. Saward represented the reformist movement in medicine, and came to that position with certain assumptions about the medical care system. Based on research that was poorly designed, such as varying patterns of practice for the same physical condition across geographic areas, he argued that most medical decisions made by physicians involved overtreatment. That research design overlooked routine biological variability, ethnic and racial variability, and value judgments about the acceptability of certain outcomes. The U.S. is a heterogeneous population and variability of that sort needs to be controlled in the research design. Value judgments about outcomes resulting from non-intervention or intervention also contaminate the quantitative surface of that body of research. Even medicine's own limitation as a determiner of mortality (and mortality is used rather than careful morbidity as a marker) means the best of medical intervention is a small signal lost in the noise of the other, major determiners of mortality. This, however, was the beginning of taking control of medical dollars away from the individual who decided on medical intervention, taking control away from insurance systems that were fee-for-service, and ultimately taking control away from physicians who once were economically-independent units. It was the start of managed care, orchestrated in academic medical schools.

This joined a list of books assuming the status of medical economics was in disrepair. Sylvia Law argued the Blue Cross insurance plans had failed in *Blue Cross: What Went Wrong?* (2). Another, *Paying More But Feeling Worse*, concluded the system was broken. Studies began comparing the percent of Gross National Product that was used up by medicine, and made the automatic assumption that these percents were too high, even comparing them to other countries around the world. The percent of GNP indicates the dollar choices being made, and the assumption was that, while dollar choices made to purchase autos generated a healthy economy and good prioritization of resources, dollar choices made to buy medical treatment did not generate the economic results of jobs, growth, and infrastructure, and were very poor choices economically. The medical industry was treated differently than other industries. It was not seen to be a product that could be exported on the global market (although in fact it can and is). It was not seen as a job-creator except for a small, wealthy class of physicians (although in many communities it may be the leading creator of jobs). It was not seen as an industry for the future (although the Human Genome Project and medical technology should have dispelled that mistake). It was seen as a user of resources (but so are all industries). The question then becomes, why was there such an unusual treatment of the medical care industry?

We can fastforward to a country starting to follow the U.S. approach to medical care, Canada. Canadian medical economics experts are beginning to talk about "integrated health care" and prospective payment schemes rather than fee-for-service and their current centralized government insurance. The key word for prospective payment plans is becoming part of their vocabulary as well, "capitation." That also, from a practical viewpoint, is familiar to me. Again, at the beginning of my work in bioethics, I worked with a professor of preventive medicine who was attempting to institute a capitation plan (by any other name) for our community. It was finally called "Integrated Mental Health." I participated in the planning stages and listened to the strategies to push it through in the face of anticipated clinical and community opposition. There was still strong opposition to replacing fee-for-service with capitation plans, and given a free choice between capitation and fee-for-service, most people in our community were still choosing fee-for-service. The man on the street knew the difference and voted economically, but the vote was not accepted by reformers. People knew that capitation did what it said, it capped the amount of medical resources available to them in any year. They may not have known that this downsizing of availability would continue over the years, that built into the plans was a downward spiral of removing medical dollars and resources from the system. In the years following, they would see small community hospitals closed, difficulty in being admitted to hospitals, shortages of medical staff, treatment options curtailed, and rationing in all its forms welcomed into the system.

But the policy experts were convinced the public was making a mistake by not embracing capitation plans. After all, they believed the average person going to a physician was more likely to go because of a medical demand rather than a medical need. They assured the public that true medical need would never be compromised, and that only medical demand would be rationed. The ethical assumption was that a capitation system would not lower quality of care, only unjustified costs and demands. How one drew a distinction between need and demand was never made clear. This is an arbitrary distinction as it stands, and it involved ethical decisions that have never been straightforwardly and fully presented and discussed with the public. Those making the choices, as Calabressi and Bobbit admit, are not elected, but appointed, and are not directly responsible to the voters in a democracy (3). In many ways, their role in constructing economic first-order decisions is to conceal the ethical choice from the public so as not to offend social conventions about ethics. The ethical discussion remains focused on economic second-order decisions. In other words, what are the ethical choices given that "x" dollars are available and the system (and those in it) must sacrifice because "x" is not enough to meet the need.

Returning to the integrated health care plan, a strategy was developed for presenting a capitation plan for psychiatric care in such a way that it would be accepted. The original plan was proposed by a health administrator's assistant. The administrator and the assistant both assumed there was overtreatment in the medical system, and both were ardent reformers. With the professor of preventive medicine, they met first with all the directors of psychiatric care facilities. These meetings met postmodern bioethics' requirement of being democratic in form, but in fact the trio of activists set the agenda, and whoever controls the agenda can manipulate any committee meeting. An appeal was made to the administrators that an integrated plan would give them a definite amount of money to allocate each year, a certain and clear budget. It would make management much more efficient and easy. There was little appeal to quality of care or improving patient care. The focus was on cost-efficiency, elimination of duplication, specialization of services, generation of "data," and avoiding state regulations if the plan were accepted. The administrator for the state facility would have a hard time getting this plan past his union leaders, but except for his managerial concern, the other administrators approved the management aspect of the plan.

Halfway through this process, I began to hope this proposal would never be approved or funded. It started as a demonstration grant, with someone from the state helping in the process, since the state had an economic interest in cutting the costs of state psychiatric care. A number of times, it failed to make it through the state legislature, but it kept being presented and eventually was approved. I am not proud of my small part in helping with this project, but loyalty to the professor I was working with and skepticism that anything I could say would make a difference kept me from openly criticizing this plan. That may have been an ethical mistake, for which I will have to pay with regret and resolution not to let something similar happen again. After the years of pushing this plan, the integrated mental health plan finally became a reality. Now managers could know at the beginning of each year how much money would be allocated. Anyone could have known, if they had wished to, that it would not be enough.

The incentive of the plan was to keep the patient out of the hospital. That fit closely with the state's goal of closing state psychiatric facilities and moving psychiatric patients into the community, where there was the illusion that services existed or that those that did could do the job. Serious and chronic psychiatric illnesses were expensive drains on the state system, it was assumed, and many patients found themselves for this and other reasons wandering community streets. Data was supposed to be accumulated to verify that quality of care was not being adversely impacted. Outcome measures included the old standards: consumer satisfaction (a poor marker for anything) and repeat hospitalization (but with the variable that the plan discouraged such repeat hospitalization rendering this marker very suspect). Outcomes did not include arrest and incarceration, the most common alternative for dealing with psychiatric patients (4). They did not include followup on loss of contact which could be due to mortality or disappearance from any care system at all. Continuity of care was never a consideration, and patients in such integrated systems receive non-integrated managed care that is spread among multiple caregivers and agencies. The case managers attempt to integrate actual practical care, often becoming the only consistent caregivers for routine needs of daily living, but lacking in clinical capabilities.

The "integration" in integrated care is an administrative one, not a clinical one. Medical care becomes more centralized and bureaucratized, with more caregivers interacting with each other at meetings, more raw data generated, more care review documents, more levels of management, and less direct contact with the patient. Such a managerially-bloated system would be more expensive that a fee-for-service system if the patient population in need of treatment actually received treatment. However, this integrated system is constructed so that it is easier for patients to become lost to the system or to become so confused moving around the system that they can no longer negotiate it. The actual amount of clinical care continues to be downsized, which is the real purpose of such delivery systems if we were honest about them.

The amount of managerial expense and data collection are higher than in the old system, adding unnecessary but predictable levels to the bureaucracy. It is beneficial to professional managers, but not beneficial to patients, physicians and other medical staff, and ultimately society.

One would expect a flurry of ethical criticism from bioethics, but this will not occur in postmodern bioethics except when the system is for-profit. When these delivery systems were non-profit, postmodern bioethics fully supported them either by its silence or its advocacy. Only when business and profit became part of the managed care, integrated care systems did some ethicists begin to criticize them for lack of quality of care, and, most importantly, for the capitalist goal of profit and greed. Such lowering of quality of care, however, has been going on since the first HMO was created, and non-profit managed care systems use exactly the same delivery system scheme for downsizing as for-profit managed care. Rationing, restricting research and development, capitation, managerial definition of "medically necessary," and requiring patients to navigate complex systems with deliberately built-in roadblocks that use wording, queuing, and repetition have all been part of the earliest integrated or managed care (5–8). The ethical issues with such managed systems were the same then as they are now. Those questions have not been asked or answered, partly because, as Wikler notes, bioethicists are not that welcome in policy decision making (9), partly because medical economics is not a specialty of ethics, and partly because bioethicists have persistently refused to recognize and own their ideological goal of cost-efficiency within delivery systems. Postmodern bioethics has not assumed its obligation to answer for what has been done to the medical care system through its legitimizing of cost-efficiency, but what has happened raises large ethical issues and demands an accountability.

THE HEALTH BELIEF MODEL AND THE CHANGE OF NAME TO HEALTH CARE

I will return to the Health Belief Model in more detail in Chapter 9, but it is also important for our discussion of the changes in the medical care delivery system (10). The Health Belief Model is intimately connected with bioethics in three ways. First, "health" involves scientific and ethical concepts in its formulation, as the Greeks knew long ago. How we evaluate variation from a norm, as sufficiently functional or malfunctional, is implicitly ethical judgment. The very words, good or adequate functioning and malfunctioning, are ethical concepts whose first careful descriptions are in Plato and Aristotle. We can make them as statistical as we wish. At the base of our statistics is humanistic, naturalistic ethics. Second, the Health Belief Model is a utilitarian ethics of the purest form. It always chooses for the group level of organization against the individual level, or assumes the individual's good is subsumed under the group good. For any particular individual, the actual health risk may be quite low for the individual. It is when those health risks are amplified at the group level that society determines they are serious risks and that the individual's behavior must therefore be curtailed. This model, then, is willing to always sacrifice the freedom of action or the sense of acceptance and pride of the individual for the commonweal. Third, "Health Belief" is actually quite a correct description, because often the definition of health is based on no more than belief. It is open to the risk of arbitrary definition for any purposes, and can irrationally and unethically supply the "good" cause or end which justifies much unethical means and method. It is easy to manipulate. It is also a powerful manipulation, because the term "health" has received an enormous status in our contemporary society, substituting for other values that have been discarded, such as honor, courage, honesty, pride, love, meaning. Anything can and is justified by an appeal to "health." This fits well with postmodern ideology. This ideology makes the choice for the public/political level of organization, the social cause, liberation theology, community welfare, public efficiency, and public praxis. It is inherently inimical to the individual level. In bioethics, this means the Health Belief Model will chose for the construction of economic first-order decisions within its political framework, and for centralization of managed care systems. It means that the goal or end of "health" will supercede all others if that goal works to further the total society. Because "health" is so undefined, it is a strategic tool with many uses. It can replace "medical" and place the emphasis on healthy people in the system, rather than people who have diseases requiring expensive medical care. It can be defined in terms of preventive care, health maintenance by the individual, health responsibility on the shoulders of the individual, and priority of groups who are keeping their health over treatment of individuals who have a disease or condition necessitating actual medical intervention. It can even justify limiting access to care or refusing care to those who have self-caused illnesses, or whose illnesses will not allow a return to a health baseline (11).

Somtimes, an argument can be made by fiction that is as powerful as one made by philosophic argument, and I think T. Coraghessan Boyle has succeeded in that with his novel, *The Road to Wellville* (12). Boyle's novel is a wonderful satire on the irrationality of health beliefs, and the real human costs of such beliefs. It is a good example of why science and good research design needs to be reintroduced into health research, and why postmodern science should make a hasty exit from that field. Early bioethics used to enjoy make fun of traditional medicine and its unscientific assumptions about behaviors such as masturbation (13). It has been conspicuously silent about the possibilities of making fun of unscientific current health care research, with uncontrolled variables running free and with ideological beliefs determining design and conclusions. It was left up to a fiction writer to do that necessary first step.

There have been a series of changes in terminology within the medical system that reflect this postmodern ethics. We no longer speak of medical care, but rather, of "health care." That may seem a trivial point in an argument, but the Chinese long ago understood that he who controls the definitions controls the argument (14). If we have a health care delivery system, we have already chosen for preventive care, no matter the lack of sound scientific verification of its beliefs. We have chosen to construct a system that emphasizes treating healthy people and keeping them within the healthy group as long as possible, whether such an activity is probable or not. The other side of this coin is that individuals born with health problems or individuals who are no longer capable of being kept in the healthy class are now the new underclass in medicine. Their care is expensive, and there will be a systems-structured choice against receiving such care. Cost-efficiency or profit can be made most easily on the class of the worried-well, and what follows also is the proliferation of new "syndromes" to accomodate treatment of the healthy. These syndromes fit a certain profile: treatment required is brief office treatment, ambulatory status is assumed, continuing brief and inexpensive treatment is preferable, cure is a moving goal, and the patient's productivity is not impacted.

We have chosen to construct a system that is difficult to access for seriously-ill individuals, and once in that system, there is pressure to accept disability and death. The number of "inefficient" community hospitals is consistently lowering, hospital mergers are increasing, and beds are rationed. Extremely short stays are common, with morbidity not a major concern. Ambulatory surgery is increasing. Nursing care is being shifted from the hospital to the home, or, on occasion, to the nursing home. It is true that organ transplants and heroic surgeries continue (with some age and lifestyle rationing), but these medical interventions are likely to return the individual to some acceptable healthy baseline, some social utility, and are accepted as cost-efficient.

Hospitals are now set-up in terms of levels of care, with tertiary-care hospitals defining themselves exactly as described above. They are for patients who stand a reasonable chance of being returned to functional baseline. Other patients who will require continuing nursing care, who are in the process of dying over some time, or who would need extensive technological support are not seen as appropriate for a tertiary-care hospital. Even differential diagnosis is problematic in such systems. A Persian Gulf War veteran in a tertiary-care facility was actually evicted by the county sheriff after refusing to be discharged. The veteran suffered from Gulf War Syndrome and in outpatient treatment was continuing to get sicker rather than better. He was hospitalized for a battery of tests, but his physicians could find nothing specific and could recommend no further treatment. To the man, that must have seemed a death sentance, since his physical condition had deteriorated badly and he and his wife felt he could not be cared for at home. He wished to remain hospitalized until some medical answer to his dilemma could be reached. The hospital had the legal right to evict such a patient who refuses to leave, and it chose to exercise that right. At one point in the systems change, such patients might have been able to access small community hospitals, which were very good at humanly caring for such cases. Now, those small hospitals are closing or emphasizing health education. The managed care systems do not make them economically viable.

Medical care has been replaced with the term "health care" in general. This allows health education, lifestyle modification, alternative therapies, and office monitoring to replace traditional medical care. The term makes no distinction between scientific medicine, shamanic therapies, behavioral manipulation, ritual office visits, or self-help education. These less-expensive therapies are not required to have careful outcome analyses or verification, and are often covered under insurance systems. They shunt a certain percent of individuals away from the more-expensive medical care and into ideologically-approved activities.

It also follows that one does not need to be a medical professional, and that "health professionals" are all equal and accepted. Lower-salaried health professionals can then attempt to deliver the care that high-salaried and extensively-trained physicians once did. The economics of this benefits the delivery systems, but on occasion also benefits the physician. An obstetrician or family medicine physician can be seeing one pregnant patient in his examining room and billing the full amount, while at the same time, his "supervised" midwife can be seeing another patient for perhaps half the amount for her and half for the physician. The midwife is generally ideologically-opposed to medication and other expensive medical intervention, and can effect a more economical delivery. There are a host of health care professions whose salaries are below those of a physician and who are allowed to perform many of the functions of a physician. That will only increase over time. Individuals who are not certified phlebotomists are drawing bloods, practical nurses are doing the activities of nurses, nurses aides may be doing the work of practical nurses, case managers are trying to accomplish what psychiatrists and psychiatric nurses did, health educators do what physicians once did – the staffing requirements of hospitals have been changed by the name change to health care professionals.

The most recent change, to "health care workers," is ethically troubling. Aides are attempting to do psychiatric therapy with inpatients, midwives need not have medical training, alternative therapists with no medical training are welcomed in hospitals. Just recently, an emergency medical technician described to me a case she had where the baby had presented at home with a full-leg breech. The parents had elected to have a midwife who, although registered as a midwife, was not a licensed nurse and who had not made it to the home in time when called. The ambulance crew managed to deliver the baby, who was blue and limp, with no neck or fontanel pulse. Two hours after mother and baby arrived at the hospital, the midwife arrived in a dirty sweatshirt, unkempt hair, and with an angry voice demanding that the mother be released and sent home, although there had been extensive tearing during delivery. The baby died a few hours later, with the midwife still complaining that the crew should have taken mother and child to the small local hospital which had no facilities for such a delivery complication, and should not have instead met the medical technician who was in his car coming toward the ambulance to link up. The parents learned nothing from this tragic experience, and had nothing but praise for the midwife and criticism for the ambulance crew who had come to the memorial service. That crew had, in fact, gone back to the home and washed the blood off the bathroom walls so that the parents would not have to return to such a traumatic scene. One of the crew members walked out of the memorial service.

Finally, what the change represents is that the profession of medicine has had its profession and its facilities taken away by postmodern bioethics and its support for current medical economics and policy making. The goal of postmodern bioethics that Rothman postulated specifically and that this book has attempted to demonstrate through specific issues analysis has been met (15). Part III will attempt to show that we are all the worse off for it, and suggest we must attempt to undo the damage.

THE PATIENT AS A BUSINESS UNIT

Joseph Fletcher was a friend of mine and inclined to listen to my concerns about the abuse of ethics that I thought postmodern bioethics represented. When I would tell him about typical cases at our hospital that illustrated the cost-efficiency nature of this bioethics, he would be critical of the case decisions. Still, he couldn't agree that they were typical or that bioethics would not be self-corrective. He underestimated the economic and ideological forces within postmodern bioethics and the changes of the 1970s. Like Ruth Macklin, he failed to see the inadequacy of the law or of personal choice in protecting patients against such systems.

Bioethicists and humanists also resisted looking at the economic pattern of abuse of ethics: its use as a strategy to weaken the authority and power of clinicians and bioscientists in order to radically change the delivery system and achieve the economic agendas of downsizing and efficiency. A bioethics analysis during grand rounds that identified this economic purpose was dismissed by assuming such analysis must be absurd or paranoid. Mad doctors weren't roaming the halls looking for patients to terminate. No, they weren't. They were sitting in medicine grand rounds accepting a new addition to each case discussion: an economic cost/benefit determination of the appropriateness of continuing medical treatment or instituting expensive treatment. Quality care had already been defined in such medical school rounds as cost/effective care, or utility/futility. Quality of Life was being introduced as a utilitarian criterion for justifying or not justifying treatment, and the quality of that life would have to be quite high to meet the standards for justifying treatment. No one was in the hospital corridors; they were all at grand rounds.

Postmodern bioethics continued to find that the system could be coercive and manipulative in situations such as patient assent to research, costly technological and experimental treatments, and care that prolonged life. At the same time, bioethics would not consider, as the circuit court of appeals justices would not consider (16), that patient choice could also be manipulated and stuctured to achieve a downsizing goal. Bioethics, because of the ideology at its core, continued to talk about overtreatment when the actual problem in the medical care system was undertreatment, at least for serious, scientifically-established diseases. There is even greater risk for the patient, however. Bioethics, even when occasionally recognizing that some patients are sacrificed in the new delivery systems that have been created, has another arbitrary principle with which to supplant the principle of autonomy, the principle of justice. "Justice" is defined in a neomarxist manner as the just redistribution of resources toward broader social goals than medical care for the seriously ill who cannot be returned to baseline. Well-baby clinics are placed in direct competition with expensive medical treatment that can perhaps benefit only a few persons, with the assumption that well-baby clinics will get not only priority in the redistribution of resources, but that such resources freed-up from expensive medical treatment will somehow find their way to clinic budgets. Of course, there is no mechanism, no channel for getting such third-party-payer savings to these clinics, and no evidence that such resources saved actually get to such clinics, but postmodernism is interested in subjective intent, not objective results. This argument is part of the unproved position that the medical care system should be downsized, and it is hidden from ethical analysis and public discussion within the appointed governmental bureaucracies that make economic first-order decisions. The ideology of bioethics supports such centralized planning for its efficiency and cost-containment, and for the end result having been achieved, taking control of the medical care system away from medical experts who were not willing to cooperate or be coopted by that ideology.

It is not possible to give a bioethics analysis of any issue in the medical care system without some understanding of current business school concepts and managerial systems. Any attempt to develop an adequate bioethics and maintain the values implicit in medicine, the biosciences, and humanism requires an understanding of the processes incompatible with those values. There has to be an identification of the risks being created in the new health care delivery systems, risks to individual patients and to society.

For example, a well-known psychiatric hospital, on the verge of bankruptcy, became a health care system, as so many facilities are now called. This signals the change into business systems funded by both public and private contracts. Hospital administrators have quietly become CEOs, without much comment from bioethics. The managerial class has taken control of hospital systems, as it has taken administrative control of most social systems. The managerial goal is to operate efficiently and effectively and to capture market shares. Most mission statements are positive jargon, such as "innovation and excellence in the delivery of behavioral health, mental illness, and addiction" for this psychiatric hospital/health care system.

The devil in the details of such mission statements is that the actual goal is: a cost-effective continuum of care to reduce cost per episode of illness; the use of prospective payment, risk sharing (the physician who made the expensive decision to treat will find his salary reflecting that loss), and capitation; the expansion of non-inpatient services (not only reducing access to hospitalization but concentrating on the money-makers of soft syndromic illnesses); the development of alliances and joint ventures (a polite way of saying this system will begin to create a vertical monopoly for health care in its community by buying up lucrative practices or forcing mergers). A vertical monopoly is similar to John D. Rockefeller's monopoly, Standard Oil. All levels of the health care system are owned by a super-board (for-profit or non-profit): community clinics, office practices, hospitals, hospices, nursing homes, respite care, rehabilitation clinics, community social agencies, transport services, medical supply companies, diagnostic laboratories, senior independent living complexes.

The health care systems manager speaks in terms of service/product lines which are business units that must capture a large share of the external market, meet the needs of third-party payers (who are its real customers), and produce a positive bottom line. These business units will be competing for private and public contracts and must grow in that area (inexpensive medical treatments), while downsizing expensive beds and treatment, and increasing staff productivity (more work for less pay).

The former hospital-now-health-care-system will rent out expensive inpatient beds while buying outpatient practices and extending into community social engineering programs which have little to do with what's left of science in psychiatry. The choice of the term "behavioral health" was not accidental. Lifestyles may be manipulated as medical treatment, and neoleft social programs will become folded into the health care system, with all the aura but none of the reality of being medical. The politicizing on this level will have achieved a segment of the total society.

Workforce productivity will be achieved through "relative value units," and salary will be annually negotiated between the managerial class and the health care professionals who are rapidly becoming health care workers. Salary will be based on "target units" (how much time the health care workers wish to invest to make a stated amount). Patient care is a phrase not seen in any of this presentation. In the health care system in question, the managerial class achieved the results of 20% fewer clinicians seeing 30% more patients. Also within this system, private practice was being phased out for those professionals, so that all would be reduced to accumulating units to maintain salary. "Units" are part of a new production line role for health care workers, while "practice" was the old professional expert role, rapidly disappearing.

Lest the reader think this is an exaggeration or science fiction or my own idiosyncratic interpretation of a more ethical process, I am merely paraphrasing the visiting administrator's grand rounds presentation that took place in our department to institute the change in our own health care system. It was an accurate and honest description of how management had changed a traditional psychiatric hospital into an efficient health care system. The steps to this state of affairs can be seen by anyone who wishes to see, and postmodern bioethics had a major role in achieving this change by:

(1) criticizing and eroding the professional expert tradition and deprofessionalizing medicine (17);
(2) labelling the Hippocratic ethical system as outdated, irrelevant and invalid, thus removing the moral basis for physician resistance, devaluing medical treatment, and giving positive reinforcement for non-treatment (18);
(3) denying scientific expertise and reality-based nosology and thereby fostering the belief in the cultural construction of disease and its conservative political biases (19);
(4) assuming a natural rather than artificial (economic first-order decisions) scarcity of resources and legitimizing the unnecessary sacrifice of individuals in order to redistribute resources (20);
(5) placing the burden of choice and consumer protection solely on the shoulders of increasingly-isolated and sick individuals who have neither the physical strength nor knowledge of the new health delivery systems to make an adequate choice (21);
(6) desensitizing the culture to death and the discarding of individuals without social utility, unless they are members of a victim group (22);
(7) disguising or denying the economic and political purposes/goals which are an integral part of postmodern bioethics;
(8) serving to carry out cost-efficiency programs through its own bureaucratic structures of ethics committees, human subjects review committees, governmental commissions, task forces, consensus institutes;
(9) criticizing only for-profit systems and not informing the public that all managed care systems operate in similar ways, including governmental ones.

THE LACK OF STATISTICS AND STUDIES

Dr. Richard Ciccone and our other co-authors may have been some of the few researchers to have the opportunity to do a study on the effects of legal and bioethics changes on patient quality of care (23, 24). This research was done without grants or funding, since it would not easily have fit into the protocols of policy makers, but it produced two good articles that inevitably would bother reformers. In New York State, the highest court in its *Rivers vs. Katz* decision (25) gave involuntarily-committed psychiatric patients the right to refuse medication, unless a review judge determined they did not understand this refusal and the medication would benefit them. These Rivers hearings constituted an expense for state and private facilities, and generated another level of administrative/judicial review in addition to the previous review. The purpose of the study was to see whether there was an impact on patient care. It was done over years prior to the Rivers decision and years following the decision. To check results, a second review of treatment refusal and medical charts was done for another year. There was, in fact, an impact on the quality of care patients received following Rivers.

It is very difficult, however, to do such studies within the entire medical care system, and the results of such studies will not be well-received by policy makers or bioethicists because the results might violate an assumption that quality of care is not impacted by downsizing the system. Yet there are court cases which demonstrate that quality of care has decreased, at least for some patients who go to court with claims (26), and there are cases that are known first-hand by some professionals which indicate the same thing. Such cases are usually dismissed as "horror stories" which are anecdotal evidence only (and therefore not acceptable) and lack good research design and statistical studies. The court cases are also a source of difficulty, because they too are anecdotal, though usually with enough evidence to convince a judge or jury. The legal cases are now being restricted even further because of a section in an old law, the ERISA clause, which the U.S. Supreme Court has ruled prevents patients from suing managed-care systems (27). With the legal avenue of redress now cut off, there will be even fewer indications that something has changed, and not for the better, in quality of care.

Medical professionals could inform the public that quality of care has been lowered through cost-efficiency and downsizing, something the public, not being quite so naive as policy makers often assume, already knows. That, however, would require admitting that the professionals are not delivering the quality care that was the community standard of practice in previous years. It would also make them subject to the argument that they are claiming this only for selfish reasons, their own self-interests making their claims unbelievable. It would also require some ethical decision on their part about allowing their behavior to be structured by a system that might be unethical. It would be the sort of crisis of conscience that postmodern bioethics could not understand or incorporate. In postmodern bioethics, the choice has already been made, assuming that conscience or objective ethics is not valid, that good people should come together and accomodate. There is no justification for angry men in postmodern bioethics, nor for passion. It is a method for incremental acceptance of whatever the consensus produces – provided, of course, that those negotiating already believe the underlying beliefs of the postmodern tripod. A professional, bioethics would insist, must accomodate to the utilitarian, cost-efficiency value, and not be concerned with whether the individual patient's sacrifice was necessary or not. All decisions must be seen as economic second-order decisions, and the important economic first-order decisions must not be identified, analyzed, or rejected. Clinicians should remain at that second-order level, trying to do the best they can within the limitations that continue to be imposed on their patients and their practice of medicine. It is predictable that they will do just that.

SHOULD HORROR STORIES COUNT

The family of a 93-year-old woman asked the New York State Department of Health to investigate the care she received at a teaching hospital, one of two cases of the care of elderly women in the emergency room that had occurred recently. In fact, these are not isolated anecdotal cases, but the extreme warning flags of a bureaucratized, postmodern system that has fundamentally changed medical values. This woman (and also the other) was taken to the emergency room and then quickly discharged because that is what generally happens in the system, and it is the system, not the individual physician, that most forcefully determines what will happen.

The woman had been picked up from her bed and taken to the hospital in her bed clothes during midwinter. This is routine for ambulances who "scoop" the person up during an emergency, drape with warm blankets, and transport as quickly as possible. Time, precious time, is lost if an attempt is made to dress the patient, and this patient was exhibiting what could have been serious problems. When she arrived at the emergency room, she was seen, evaluated, determined not to need hospitalization, and put in a taxi to be returned home in the late hours of the evening. She had only a thin hospital blanket to protect her from the cold, and hospital paper slippers. She was apparently exposed to the cold for some time. She remained at home only a brief time before symptoms reappeared, and she was taken back to the hospital. This "revolving door" experience is typical for many more patients than it should be. This time, the woman was admitted, and 32 hours after the initial discharge, she died in the hospital of an apparent heart attack.

The hospital regretted the premature discharge and expressed sympathy to the family. However, the hospital also stressed that this premature discharge did not seem to be the cause of her heart attack, a conclusion that the state had also tentatively reached. The hospital medical director did make a courageous stand by telling the public that emergency room facilities are stretched too far and the system is underfunded, one brave but small crack in the wall of denial that quality of medical care has not suffered from the cost-cutting and downsizing of the new health care delivery systems.

The State Department of Health had to take some action, and made recommendations about proper discharge of patients who are brought in without adequate clothing. It was suggested that clothing be checked and the hospital provide clothing before discharging a patient to the taxi and the street. It was also recommended to ambulance crews that they make sure their patient being transported had shoes and clothing to suit the weather, although crews do not feel this additional time is safe to take in many circumstances. In a life-threatening emergency, it would be unethical. What this suggestion accomplishes is a shift of responsibility from the hospital to the ambulance crew, and is not ethically warranted. What the State's recommendations do not address sufficiently is the organization of the hospital health care delivery system that would allow such treatment of a 93-year-old woman. The professionals within such a downsized system will be structured to act unethically whether they wish to or not. It is the Exxon Valdez all over again. There, the economic first-order decisions about oil transport create a system where decent people will find themselves taking risks as the captain of the Valdez did in trying to avoid iceberg-clogged water in a single-hull tanker in the middle of the night without Coast Guard escort and without an adequately-trained third mate. The easy approach was to point to Captain Joseph Hazelwood's problem with alcohol and the fact he had been drinking in a bar in Valdez before boarding the ship. That misses the entire ethical problem with the economic decisions made in the system. If Captain Hazelwood had suddenly become ill and had to leave the bridge for that reason, the same series of events would have taken place as if he had been intoxicated and had to leave for that reason. He was in exactly the same position as the captain of another tanker off the coast of Spain who had waited days for a berth for his oil tanker. Desperate about the delays and the economic pressures, he decided to attempt to bring the ship in at night in the middle of a storm, and ran her aground just as the Exxon Valdez ran aground (28). Both men were structured to act as they did by the economic first-order decisions within the oil transport delivery system, a system which is basically unchanged because those first-order decisions do not receive public debate nor depend on public approval of the risks. Physicians within our health care delivery system are structured precisely in the same manner as Captain Hazelwood, which is why a very elderly woman was sent out into the winter cold, basically to live or perish on her own.

We are faced with a convincing conflict between classic bioethics and current postmodern bioethics. In the current bioethics, the following assumptions or value judgments are made: Cases like the old woman's are so rare and anecdotal that they should not carry the weight of a proper research study. These patients were very elderly and likely would have died no matter what care was given, so it was not cost-efficient to care for them beyond a quick evaluation. These patients had multiple medical problems and did not die as a result of the presenting complaint or premature discharge from the emergency room, so there is no ethical issue. It is the family's responsibility to care for their elderly, not society's. It is better to die at home than in the cold and technological hospital. The emergency rooms had to use triage because of unusually high use and we cannot afford any flexibility in the system. It will be necessary to do more rationing of health care because of lack of resources, and treating such elderly patients must be sacrificed to the general good.

These postmodern assumptions collide with a different set of values, the values of classic bioethics and humanism: Persons afraid they are in a medical crisis come to hospitals for help, expect to receive it, and should receive it. Families bring loved ones to hospitals themselves if possible, or transport if not, and expect the needs and interests of their loved one to motivate the decisions made there. A thorough differential diagnosis takes time and costs money, but it is the gold standard, along with continuity of care, for quality care. Chronic care and acute care are both the mission of a hospital, and most health care is eventually chronic care. No one has asked the patient or the physician to vote on economic first-order decisions that remove money from the medical care system, a vote should be required, and if it were, most persons would not vote for the present funding limits or the cost-efficiency values that follow from such limits. People do not expect to get thrown out of hospitals or prematurely discharged, nor should they be. The feelings and fears of women in their eighties and nineties count ethically.

These two sets represent the value conflict embedded in the medical care system and also in social systems as a whole. Classic bioethics was Aristotelian and Roman at its base, defending the innate worth of every human being and critical of the means used to achieve even good ends. Postmodern bioethics is at most utilitarian, but only in an economic way that would horrify Bentham and Mill, willing to justify any means for a postulated good end, and requiring human worth be proved cost-efficient. It is a denial of the importance of content-ethics, and hence ethics, and the assertion of the sovereignty of political institutions. What chance would two elderly women have against such a system?

HOPE AND UTILITY

Hospitals were never a "safe" place, although the Knights Hospitallers tried to make them as comfortable and hopeful places as possible, even in the middle of battles during the crusades. Within religious traditions, they were the House of God. Even after the religious identification had passed, they were places of hope. Hope counts – or at least it once did. Hope, however, is not cost-efficient. Still, taking hope away from a 93-year-old woman and replacing it with fear and abandonment is not a value of the medical care system we once had. What is economically happening in the new delivery systems is a refusal to accept the costs of hope, a denial of the worth of those who no longer have social utility, and the requiring of human sacrifices. I mean this quite literally. Cost-containment, futility/utility, and cost-efficiency are less-bloody means of human sacrifice, less open but just as thorough. Human beings can be killed by economic tools, can be no longer economically viable, and just as dead as by sword or fire or gas.

Within objective ethics, there is sometimes a justification for utilitarian sacrifice, but it requires a great deal of justification, well-argued. The fundamental question, not asked in economic first-order decisions, is: Is this sacrifice necessary? I have argued for some sacrifice in some situations, but I also argue for appreciating the enormity of what is being done. I can think of no better ways to put it than a book on the history of the devil in antiquity by J. B. Russell (29):

> "The story of the Devil is grim, and any world view that ignores or denies the existential horror of evil is an illusion. Ivan's one child crying out alone in the darkness is worth the whole creation, is in a sense, the whole creation. If any world view, theist or atheist, minimizes her suffering, declares it non-existent, gives it elaborate philosophical justification or explains it in terms of a greater good, whether the name assigned that good be God or the People, that world view renders her life, and yours, empty and vain."

I can also think of no better description of the unethical economic first-order decisions basic to postmodern bioethics.

REFERENCES

1. Potter, V. R. (1971). *Bioethics: Bridge to the Future*. Englewood Cliffs, NJ: Prentice-Hall.
2. Law, S. A. (1976). *Blue Cross: What Went Wrong*. New Haven, CT: Yale University Press.
3. Calabresi, G., & Bobbit, P. (1978). *Tragic choices: The Conflicts Society Confronts in the Allocation of Tragically Scarce Resources*. New York: Norton.
4. Ciccone, J. R., & Clements, C. D. (1984). The ethical practice of forensic psychiatry: A view from the trenches. *Bulletin American Academy Psychiatry and Law, 12*(3), 263–277.
5. Franks, P., Clancy, C. M., & Nutting, P. A. (1992). Gatekeeping revisted – Protecting patients from overtreatment. *New England Journal of Medicine, 327*(6), 424–449.
6. Jecker, N. S. (1990). Integrating medical ethics with normative theory: Patient advocacy and social responsibility. *Theoretical Medicine, 11*, 125–139.
7. Jonsen, A. R., & Jameton, A. L. (1977). Social and political responsibilities of physicians. *Journal of Medicine and Philosophy, 2*, 376–400.
8. Aaron, J., & Schwartz, W. B. (1990). Rationing health care: The choice before us. *Science, 247*, 418–422.

9. Wikler, D. (1994). Bioethics in health policy: What methodology? In: E. Bittar & N. Bittar (Eds), *Principles of Medical Biology*, Vol. 1A (pp. 23–36). Stamford, CT: JAI Press.
10. Jeffery, R. W. (1989). Risk behaviors and health: Contrasting individual and population perspectives. *American Psychol, 44*, 1194–1202.
11. Sider, R. C., & Clements, C. D. (1987). Medical caring for the smoker: Ethical responsibility works both ways. *Chest, 91*, 156–158.
12. Boyle, T. C. (1993). *The Road to Wellville*. New York: Penguin Books.
13. Engelhardt, Jr., H. T. (1978). The disease of masturbation: Values and the concept of disease. In: T. L. Beauchamp & L. Wlaters (Eds), *Contemporary Issues in Bioethics* (pp. 109–113). Belmont, CA: Wadsworth Publishing.
14. Kahane, H., & Cavender, N. (1998). *Logic and Contemporary Rhetoric: The Use of Reason in Everyday Life*. New York: Wadsworth.
15. Rothman, D. J. (1991). *Strangers at the Bedside: A History of How Law and Bioethics Transformed Medical Decision Making*. New York: Basic Books.
16. Quill, Klagsbrun, & Grossman v. Vacco, Pataki, & Morgenthau. U.S. Court of Appeals for the Second Circuit. (1996). U.S. App. Lexis 6215.
17. Rothman, D. J. (1991). *Strangers at the Bedside: A History of How Law and Bioethics Transformed Medical Decision Making*. New York: Basic Books.
18. Veatch, R. M. (1981). *A Theory of Medical Ethics*. New York: Basic Books.
19. Sedgwick, P. (1976). Concepts of health and disease. *Journal of Medicine and Philosophy, 1*, Special Issue.
20. Calabresi, G., & Bobbit, P. (1978). *Tragic choices: The Conflicts Society Confronts in the Allocation of Tragically Scarce Resources*. New York: Norton.
21. Culver, C. M., & Gert, B. (1982). *Philosophy in Medicine*. New York: Oxford University Press.
22. Battin, M. P. (1983). The least worst death. *Hastings Center Report, 13*, 13–16.
23. Ciccone, J. R., Tokoli, J. F., Clements, C. D., & Gift, T. E. (1990). Right to refuse treatment: Impact of Rivers v. Katz. *Bulletin American Academy Psychiatry and Law, 18*(2), 203–235.
24. Ciccone, J. R., Tokoli, J. F., Gift, T. E., & Clements, C. D. (1993). Medication refusal; Judicial activism: A re-examination of the effects of the Rivers decision. *Hospital and Community Psychiatry, 441*, 555–560.
25. Rivers v. Katz, N.Y.C.T. App. June 10, 1986.
26. Hoge, S. K. (1995). ERISA, Part 1: Barriers to regulating health care, and ERISA, Part 2: The Malpractice liability of managed care companies: A ray of hope. *Journal of Practical Psychiatry and Behavioral Health*.
27. Herdrick v. Pegram. 2000. U.S.S. Ct.
28. Russell, J. B. (1977). *The Devil: Perceptions of Evil from Antiquity to Primitive Christianity*. Ithaca, NY: Cornell University Press.

9. ALTERNATIVE MEDICINE, NEW AGE, CLASSIC SHAMAN SYSTEMS

A SHAMAN GREAT-GRANDMOTHER AND SCIENTIFIC MEDICINE

Despite the nihilistic epistemology of postmodern bioethics, there are good reasons, good philosophic arguments to choose the scientific method as the best tool to understand the reality that lies beneath human experience, and to make modifications to better the human condition. Testing ideas for the feedback of reality, and meeting standards that are designed to eliminate wishful thinking, unproved assumptions, and ideologies are the best chance to get things straight. There is no transcendental armchair or shortcut to using our central nervous system to help us exist and comprehend a segment of the cosmos. The postmodernist attack on science which demeans it by placing it alongside narrative stories and assuming it is no better than any of the other stories leads too often to tales idiots tell, full of that ideological sound and fury that Shakespeare perceived signified nothing. Our learning program, our central nervous system, is better than that. Our transaction with the world is more pragmatic than that. We do not have to be satisfied with tales or stories.

It is more than just a story that if we jump from the Empire State Building we will squish ourselves on the reality of the concrete, cars, awnings or pedestrians below. We can tell all the stories we wish about levitation, but it is not likely to happen. We are not looking for absolutes, of course. Unusual events do occur. One would-be suicide jumped off the Empire State Building, got caught in the updraft, and was blown into a window two floors below. It is only highly probable that a suicide attempt from that building will succeed. Inductively, however, we had better consider that jumping will flatten the person on the pavement because that is what is highly probable in this world. Probability is more than adequate to what we have to learn and do, and probable knowledge is superior to narrative knowledge if we follow our genetic learning program for survival in the natural world. It is the choice for sentience that life-forms with complex organization have made in the evolutionary randomness (or perhaps pattern) of the cosmos.

What does epistemology and metaphysics have to do with bioethics? It turns out, a great deal. Postmodern bioethics may relegate both to the intricacies of academics, but it does so by making its own assumptions about epistemology and metaphysics as we have seen. Those assumptions do not hold up as well for our world of experience as the scientific method does in its full reductive and wholistic range, nor are they assumptions based on valid conclusions about epistemology and metaphysics. The assumptions are ideological, not philosophical, and return us to a time before philosophy and science. We need not take that journey. Systems theory supplies part of the metaphysics and the most fruitful epistemology to counter the nihilism about knowledge that is implied in postmodernism. In Chapter 11, I will look more closely at this alternative to pre-classic darkness. In this chapter, I want to look at ancient alternatives to scientific medicine and postmodern assumptions about those ancient alternatives.

Alternative medicine may not meet the standards of the scientific method, and it would be a mistake to change those standards to accomodate alternative medicine. I can remember feeling very embarrassed sitting in grand rounds and hearing a Harvard Medical School faculty member uncritically describing chi-gong, an ancient Chinese healing by focusing the body's energy or chi. He had a slide of a chi master moving a red-tasseled Chinese lantern by mysterious telekinetic powers, and for him, that was adequate proof of the validity of chi. However, a magician could have told him how it was done, by surreptitiously blowing jets of air out of the corner of the mouth. These lanterns are very light, the chi master on the slide could easily have enough air, and the red tassels would gently sway. That same physician conducted a study of the effects of chi on patients terminal with oat cell carcinoma. All the patients died, since this is a carcinoma that is almost always or always fatal. However, the physician thought the patients visited by the chi master, who threw his force at them when he entered their rooms, had lived a more peaceful time before their deaths. This study did not use double-blind safeguards. It did not correct for subjective reporting by patients who knew the chi master had entered their room to help them in some way. It was, nevertheless, taken seriously in the halls of medicine. This is not to assert that the biological organism does not have a field of energy or that biochemical pathways are not involved in biological rhythms, or that use of scientific reduction would not produce verification of the trial-and-error empiricism of past knowledge, including Chinese knowledge called chi. In fact, one could make a good argument that the early scientists who developed these classic bodies of knowledge were using, for their time period, the scientific method, and that there are likely to be significant hypotheses we should look at from all the accumulated knowledge of our history. It *is* to asssert that researchers can be gullible or want wishes fulfilled, and that the safeguards against this that the scientific method and scientific reduction supply are critical. Without them, there will be charlatans and nonsense, and unethical harm.

I also watched as a class of medical students received a lecture from a Tuscarora Indian who claimed to be a shaman. The class listened attentively, and easily allowed the shaman to suggest to them a hypnotic state. Even when the shaman claimed that he had stopped an Arab-Israeli war by placing his hands on a map of the Mideast and willing there be peace, the students did not ask critical questions. They were quite taken by the Indian shaman. By this time, postmodern bioethics had introduced a humanities concept that it was progressive to approve of such traditions, and was calling this traditional medicine. It was also ideologically correct in terms of the victim groups of postmodernism, since American Indians are part of that entitled class and their narratives are not to be criticized but accepted in preference to narratives of the Enlightenment. This again placed me in a conflicted position, because there are indications that my great-grandmother, who lied about her baptism, may have closer connections to the Buffalo Creek Seneca Reservation than just the geographic fact that her farm was in the middle of it. An expert on the Senecas insisted to me that he was certain my family was partly Seneca and that he could tell from skull and eye features. If either Seneca or Scythian, most of us have some genetic link with the old shamanic tradition and need not be so uncritical about current practitioners who have their own private agendas.

That term "traditional medicine" has now been changed to alternative medicine, in a similar way, perhaps, as we now call the eco-correct technologies "alternative technology." Given the 1960s and 1970s name for ideologically-approved technologies, "appropriate technology," it is possible that we could have called this healing appropriate medicine.

This trendy fashion in medical schools has introduced such healing therapies not only into clinical practice but into the National Institutes of Health. This fashion is consistent with the postmodern entitlement of victims such as native peoples or developing countries, and its enmity with the standards of reason and scientific reduction. The metaphysical assumptions of postmodernism are assumed to be the assumptions of the healing traditions, but there is good reason to think this is not historically accurate. In Part I, we saw that throughout the ancient and classic world, healers were scientific botanists, pharmacists, anatomists, diagnosticians, and surgeons who could use horsehair sutures (1). The Roman army had army physicians who were quite good at treating external wounds, removing projectiles, and practising infection control (2). The Egyptians could do a case report and differential diagnosis almost as well as we can (3). Alexander the Great's Greek physician did lung surgery on him, and Alexander survived the surgery (4). The ancient healers were more scientifically-driven than we give them credit. After all, the risk to the physician/healer who failed to heal was often his own personal death.

I am not as personally conflicted in my science and my family history, then, because there is no reason to assume that healers of the past would not have been scientific, at least in part of their methodology and practice. My great-grandmother was an herb doctor and perhaps shaman, and a very strange little woman as most shamans are. I can relate to the old knowledge of herbal medication, based on that genetic learning program, and also to the strangeness of my family history. I operate with the logical standard that the person making a claim has the burden of proof to show the rest of us that the claim is valid, but I think my great-grandmother might have been able to do that given the technology that has developed since. I require adequate and valid reasons for accepting claims and few claims of alternative medicine meet that standard, and at the same time, I understand on a philosophic level the incompleteness of our experiences, the limited segment of the cosmos that we work at knowing, and the probability that there will be new scientific knowledge and theories for the future. Science is a work in progress, with the best standards and most workable means for understanding experience, and because of the nature of systems, a mystery about incompleteness that should generate awe and happiness.

My family history illustrates two aspects of alternative healing approaches throughout history. One is the naturalistic, bontanical, quite scientific for its time period herbal and biological approach. The other is the mystical (perhaps), levels of sentient beings, and design of the cosmos approach. My great-grandmother embodied the Ebenezer Society, the healing pilgrimage church of the Fourteen Holy Helpers, Indian Church and the sacred lands of the Seneca, and herbal medications Mary Kate Pfeiffer lived on farmland that the Ebenezers had sold. The Ebenezers went by many names. In Germany, they called themselves the Society of True Inspiration, but when they emigrated to Western New York they called themselves the Ebenezer Society and purchased the land of the Seneca Nation. The Society lived collectively for economic goals, but had separate homes and families. They built a communal meeting house on the site of the old Indian church, and refused to send their children to the public schools. They believed they were prophets and healers, in the context of the remnants of the Radical Reformation in Europe.

When they moved to the frontier, they sold their buildings and land to German Catholic farmers who also believed very strongly in faith healing. The Catholics erected their church on the remains of the Ebenezer meeting house and called it Fourteen Holy Helpers. These were fourteen very ancient saints, some from the third century in the Eastern center of the Roman Empire. Medieval fears of plagues and disease had combined these fourteen healing saints into a community of healers, and each of the saints was a specialist in healing different parts of the body. The mystical element was clear. An apparition had appeared to a young German boy, in the guise of fourteen children and a Christ Child. He was asked to build a church at the site, and eventually a healing basilica was constructed. When the German farmers came to Western New York, they brought their healing beliefs with them, painted a scene with all the fourteen saints at the back of the altar, began to practice faith healing, and dedicated the church to the cult of St. George (5). As a child, I recall seeing the wax symbols of healed body parts hanging on a back wall of the church, because the church became well-known as a pilgrimage church of healing. That was my great-grandmother's church, and in its small cemetery is a large stone monument for the family plot, a woman in the flowing toga of ancient Rome, with a bracelet of roses around her wrist and a classic face framed by long stone curls. This church has always seemed to have mysterious healing associated with it. In addition to the cult of St. George (another "bright son of the morning" and bearer of light in mythic traditions), there is now a shrine to the Infant of Prague added on to the church. The shamanic tradition is strong here, with roots that go back to old intellectual sources.

The other aspect of my great-grandmother's healing was the herbal. She treated neighbors for acne, and her medication was known in the area to be successful, but here we run into problems with research design, especially with a hormonal reaction like acne. She was perceived as being successful by her neighbors, at a time when herbal medication had fallen out of favor. The upper floor of her house in town was converted into sick rooms for patients she felt she should keep overnight. One man had an inflamed arm that his physician wished to amputate to save his life. Mary Kate thought she could save the arm and treated him with a special poultice for three days. The reddened area of skin began to shrink and the streak retreated. He went home. She was also reported to have saved the life of a member of an influential family from the city. All this eventually led the medical society to charge she was practicing quackery. She was not prosecuted because the community and some of its powerful members did not want her to be, having felt they had received some advantage from her treatment. She could not, however, save her oldest son from tertiary syphilis, and neither she nor scientific medicine could save most of her grandchildren from the dementia that finally killed them. There were limits to herbal medication that did not offer the hope that scientific medicine can for the future. New research on the dementias will soon open the door to adequate-enough understanding to begin meaningful medical intervention. Herbals are not capable of doing that. It is the scientific reduction in systems theory that holds that key, as we move into a fuller understanding of the molecular biological level of organization. Herbals may be an important part of the learning program when scientifically-reduced to that level, but they are a small part of the theoretical synthesis that has given hope for the future.

So there was my great-grandmother, the shamanic healer who practiced both aspects of that tradition, the botanical and the mystic. St. George and the other Helpers are assumed to be operating on another level of organization or experience, but capable of interceding at the human level. That is philosophically problematic, because it is a claim that does not have sufficient proof to accept it. There is no need to disprove it, since it is the person or institution making the claim that must provide the proof, the burden being on them. There is also no good rational reason to deny it might someday be demonstrated to be true. That is the open agenda of science, and all hypotheses can be entertained within the scientific method. Not all, however, can be demonstrated. My great-grandmother was also not like one of the Helper saints. Like most shamans, there was a dark side to her.

IMPLICATIONS OF THE BURDEN OF PROOF

At a conference on Rational Alternative Medicine (6), the question was asked about the danger of lack of scientific validity of findings and of deregulation of alternative therapies. The director of Stanford University's Alternative Medicine course presented a 10-point checklist for physicians to assist them in evaluating the research done on alternative medicine and in avoiding "junk science." Dr. Wallace Sampson's list follows:

(1) Adequate blinding when necessary
(2) Randomization procedure specified beforehand
(3) Stratification adequacy
(4) End point specified beforehand
(5) Subgroups specified beforehand
(6) Data analysis method specified beforehand
(7) Adequate numbers for power and significance
(8) Conclusions valid
(9) All patients accounted for
(10) All significant data reported.

Dr. Sampson also presented a list of "lawsuit herbs," those whose potentially toxic or harmful effects could constitute damage to the patient and a tort suit (7). Sassafras can induce vomiting, stupor and paralysis, and one teaspoon can kill an adult. Pokeroot has gastrointestinal, neurologic, and hematologic toxicity and could be fatal to children. Mistletoe may be fatal to children in as low a dose as 1–2 berries. Ephedra (Ma huang) is an amphetamine analog with its associated problems. Lobelia causes nausea, vomiting, diarrhea, tremours, dizziness, and is declared poisonous by the FDA. Life root is hepatoxic and carcinogenic. Laetrile is potentially poisonous. Comfrey is also carcinogenic, hepatoxic, and genotoxic in addition, and is secreted in breast milk. Coltsfoot is carincogenic, hepatoxic, genotoxic and a cardiopulmonary stimulant. Chaparral is hepatoxic, tumour trophic and causes cystic renal disease. Feverfew, along with many other herbals, inhibits platelet activity and is contraindicated with anticoagulants. Dr. Willem Betz, professor of medicine at Brussels University, described the Chinese herb, aristolochia, as toxic and responsible for an outbreak in Belgium of "Chinese herb nephropathy," which has also been seen in France, Japan, and the U.S. (8). Many herbals either amplify the anticoagulating drugs or depress them, and the list is lengthy. Dr. William Jarvis, executive director of the U.S. National Council Against Health Fraud, reported on data from Norway that indicated adverse effects associated with acupuncture are significantly underreported (9). If a "natural" product is effective, it is likely to have unwanted side effects that come along with it, again if we remember that the human body is a system. These products are not regulated and often have not gone through the safety and small trial effectiveness process before being available to large numbers of people.That they have been in use for extremely long periods of time is no guarantee of safety. The Europoid population north of the Tarim Basin in central Eurasia was using ephedra, hemp and poppy 6,000 years ago (10). The hemp may be relatively safe. The ephedra and opium poppy were not safe then and aren't safe now.

With herbals we are talking about products/medicines that at least have some effect, or assumed pathway of effect, on diseases. Those diseases or traumas tended to be the past knowledge of our world culture and reflect a previous understanding of the human body. There were medications for wounds, since this was a common need throughout history. Although the infectious process was not understood at the requisite level, there were trial-and-error medications for wound infections as well. The common infectious diseases were represented, something called ague or other archaic phrases like pox. Skin irritations and infections were also common. Pneumonia and heart problems were partially understood, and the foxglove plant was certainly a major medication, although since different leaves could contain varying amounts of the digitalis that helped the heart failure and standardization was difficult, one could also die from the treatment through undertreatment or overtreatment.

What is interesting is how postmodern bioethics and other postmodern trends have incorporated healing therapies. The selection tends to be in favor of those identified victim groups who therefore have special, more-than-equal status in the postmodern conversation. Chinese healing is preferred to Roman healing, because after all, Romans were the oppressors of the non-Causasians and patriarchal in addition. Greek healers take second-place to more exotic Egyptians and other African healers. Scythian healers are under-represented in relations to North and South American Indian healers, because Scythians were usually those blonde-haired barbarians who settled by force in Europe and whose Indo-European language mates conquered and oppressed the Dravidians of the Indian subcontinent, even though the Scythians were more than familiar with ephedra and suturing. Some European healing therapies do come in with the feminist aspect of the ideology, since women herb doctors were the oppressed of the sixteenth, seventeenth, and eighteenth centuries and often burned as witches. However, little of the history of healing and its relationship with alchemy and the renaissance, with its rediscovery of classic knowledge, with the fascinating career of Dr. Michel Nostradamus who walked the line between scientist and sorcerers very carefully, comes into postmodern or New Age promotion of alternative healing therapies 11). Those healers were often educated European men, and their oppressor alternative therapies have not done well in our postmodern culture. My great-grandmother's pilgrimage church and its fourteen healers are not part of the postmodern consensus, I suspect because they are Greco-Roman healers from the Roman Empire and therefore unacceptable to the ideology of bioethics. It is a studied and not-so-benign neglect, because it has its roots in discrimination, political suppression, and neoleft goals. One can be a member of WICCA practicing herbal medicine and spells, and be enfolded in the postmodern consensus. One cannot be a pilgrim praying at the church of the Fourteen Holy Helpers because that is too specific a "spirituality" which is not part of the approved politicized religious institutions.

There is another group of healing therapies associated with the mystical aspect of shamanism that is not particularly effective but does not seem to carry the risk of serious harmful effects. Healing touch therapies which are supposed to correct energy field disturbances are taught in many nursing schools and 42,000 nurses in the U.S. practice therapeutic touch (12). Therapeutic touch practitioners have no empirical means of testing their therapy, because they insist that it is not possible currently to measure energy flows objectively. This removes the requirement of establishing efficacy or outcomes. Naturopathy is an alternative healing that is so eclectic it makes a description of the methods and a careful testing impossible (13). Practitioners do focus on the rules of nature, and bringing the ill person back into balance with nature, rather than treating a specific problem. This postmodern belief in the perfect balance of nature is a false belief, but nevertheless very powerfully held, and it combines with the mystic element in shamanism to effect or correct a balance in the cosmos. Homeopathy is a therapy that embraces the dilution of an active ingredient, but it was a therapy postulated before the advances in molecular biology (14). The dilution is so extreme that worry about side effects is minimal. Acupuncture also seems to carry little risk, provided disposable needles are used. If needles are not disposable, then blood-borne infections are always a possible side effect. Boyle's book mentions a host of other healing therapies, centered on diet and bowel function, and we have not dispensed with high colonics (15). There are still mud baths, sulphur spas, relaxation techniques, imaging techniques, angel therapies, crystal therapies, roots – all having the responsibility to establish claims using the standards rejected by postmodernism: reason and empirical reduction to the next lower level (science). One might argue that even if this responsibility is not met, and it usually is not, there is no ethical issue for medicine here because these are harmless therapies. In addition, they are inexpensive therapies and sidetrack the worried-well and mildly-ill from accessing the more expensive medical therapies.

There are two major ethical problems with this. First, there is an additional obligation then placed on those still following the standards of the genetic learning program to preserve the species and to achieve some good for man. They have now been additionally handicapped in their required task, and the chances of success have been reduced for all. It is much the same problem faced by Tolstoi's Grand Inquisitor. Placing such therapies in medical schools and national institutes erodes the standards for all therapies, including what is left of the scientifically-verified ones. Dr. Marcia Angell, executive editor of the *New England Journal of Medicine* reported at a conference that 118 of the 120 U.S. medical schools teach courses in alternative therapies, and a recent survey of Canadian medical schools showed the same high level of involvement (16) Dr. Wallace Sampson also did a survey asking U.S. medical schools how such courses on alternative therapies were taught. Of 50 schools, 45 offer courses that are uncritical presentations and in his view amount to "indoctrination" into alternative ways of seeing the world.. Second, it is not necessarily true that a directly harmless therapy will not have severely harmful indirect results. Those choosing a healing therapy that is not effective have reduced their chances for healing and survival considerably. Those funding such ineffective therapies have the ethical obligation to justify the prioritizing they do with resources.

Our area has joined others around the country in setting up a treatment center for alternative medicine, already a $20 billion part of the economy in the U.S. and growing. I wanted to consider what my great-grandmother might have done were she alive today. She might have laughed to hear that this was a progressive and integrative movement in medicine and that alternative therapists were part of the treatment team. She might have shaken her head at the list of therapies practiced. What would she say about angels and crystals? Being the manager of a farm, a real estate entrepreneur and owner of her own small clinic in her home, I think she would have been practical and straightforward. Souls would have been left to the churches, separated from the human things that she and physicians could do here on earth. She would probably have used her European folklore herbs and her American Indian botanicals, but she would have judged their effectiveness by actual results in the real world.

Living outside the psychoanalytic leg of the postmodern tripod and its influences, she would have looked for an empirical cause in the physical mechanisms of the body, and left the mind to carry out its executive decisions. If she believed in faith healing, it would have been within the fold of religion, not medicine, but both might have come together in her shamanism. She didn't aspire to be a priest and wouldn't have charged money for carrying out the role of a priest. However, being practical, she probably would have requested admission to the health care team and made a very good, if perhaps unethical, living.

Why would I argue that it would be unethical? The argument would be made for the two reasons listed above. The value of the genetic learning program is a human value, and one of the bases for truth, honesty, and accumulating world knowledge. One could consider that such a value is instrumental rather than intrinsic, a means toward a good. For some human beings, however, it is an intrinsic good in itself, one of the pleasures possible for a human nature that is partly cognitive. Philosophy might even recognize it as one of the defining elements of human nature, tied to our worth and esteem. It is partly what we are. If we choose to distort that learning program, we are in a real sense doing damage to ourselves, and a humanist ethics would argue that we have an ethical obligation to our self as much as to others. It is not a contract obligation. It is an obligation of love and respect, and is more powerful and fundamental than a contract obligation. Postmodern bioethics has wanted to argue that we have no ethical obligation to our self, that self-harm and self-destruction, if others are not harmed, is an ethical choice (17). Dr. Roger Sider and I have argued that a person does have an ethical obligation to himself or herself, that ethics is not narrowly contractual or rights-oriented, not, in fact, political (18). One cannot remove oneself from the ethical community, even by choice. The damage done to a learning program that fundamentally defines who we are is serious ethical harm and needs to be recognized as such. It would not necessarily mean that illusion, denial of reality, or narrative fiction might not sometimes be ethically justified when the harm done by learning is too great to bear for an individual human being. It does mean that the cost is considerable, and is a cost to all of us who are part of the human ethical community and caught up in the human condition.

There are sufficient harmful effects from closing out the possibility of understanding natural experience through the learning program that we need to consider those effects as an ethical as well as epistemological issue. Using postmodern bioethics to challenge and diminish the biosciences and biomedicine, not because the reason or empiricism of those fields of knowledge is wrong but because it does not suit the political ideology, represents one of the evils of the total society, and a very old evil at that. When social conventions determine our science, we have treated our nature with profound disrespect. When this is done through coercion, violence, we identify a tyrannical state and a tyrant. When this is done through intellectual and political/cultural coercion, we identify a totalitarian state and a total society.

MISTAKES ON BOTH SIDES

It becomes more difficult to hold the standards of science in medicine when even scientific medicine has become so politicized that it frequently violates them. It is not just outright scientific fraud (which is violated in biomedical research out of proportion to the other specialties of scientific research) that is the issue. Within the body of biomedical research are too many examples of poor research design, manipulation of statistics, unwarranted conclusions, uncontrolled variables, retrospective rather than prospective studies, skewed samples, self-selecting samples, questionable survey methodologies, ad hoc assumptions to save the hypotheses, and the belief that collating a large number of flawed studies with small "Ns" will somehow give verifiability.

Our Prenatal Detection Program and the Human Genetics Division were always concerned at diagnostic conferences about the problems with syndromic disease classifications. Syndromes are collections of superficial or ordinary observational level clusters of signs and symptoms. These clusters have not yet been understood with a scientific reduction to the molecular level which allows identification of etiology and understanding of the progression of the syndome. As a result, they often contain multiple diseases appearing as one syndrome. They may also simply be a cluster of normal functions seen in a disease framework. Syndromes are always scientific classification problems, and our geneticists attempted to use the scientific reduction to eliminate as many syndromic disease classifications as possible. Syndromes may also be subjectively created, given their observational context.

This is nowhere as apparent as in psychiatric disease classification, and the reform of the Diagnostic and Statistical Manual (DSM) that came about in the 1970s and 1980s. The impetus for this reform of classification schemes was two-fold. There is a general problem in psychiatry with classification, because the scientific reduction is so difficult to do and the etiology is then frequently unclear. Psychoanalytic theories of etiology were not meeting scientific criteria, for example, the schizophrenogenic mother as the etiology proposed for schizophrenia. Schizophrenia historically is a good example as well of the problem with a disease sydrome being more than one disease. It once included under its too-broad umbrella all the dementias, as well as the psychiatric thought-disorder disease we now diagnose. The second impetus for reform was the perceived need for more scientifically-acceptable research. With disease diagnosis being so unreliable from culture to culture and even from psychiatrist to psychiatrist, it was felt that reliability should take preference over validity. The DSMs are now constructed so that there is inter-rater reliability in diagnosis, or appears to be. With a focus on reliability, etiology and progression of the disease took second place. Once etiology is downplayed, the scientific reduction which is the basis of etiology is also made partially irrelevant. As a result, a host of new syndromic diseases were suggested for inclusion in the DSMs, and the method of accepting these syndromes was psychiatric reliability (a majority of clinicians saw these syndromes in their practices, were comfortable diagnosing these syndromes, and felt the syndromes should be included in the DSM).

Medicine in general is no longer as careful as it should be about disease classification, and that classification can often contain an element of ideological advocacy. Advocates for battered and sexually-abused women have put together syndromes of common and minor human malfunctioning to create special diseases for such women, which can then be used in legal advocacy and political change (19). Lifestyle advocates and advocates for patients with particular diseases have supported changes in classification which place more individuals within disease classifications. The reasoning is similar to this: "Some studies indicated cholesterol played a role in cardiac disease. Medicine will then create a quantitative standard for high cholesterol which is a symptom/cause of cardiac disease. Physicians will treat office patients to lower cholesterol levels to below that standard. Treated patients achieve that lowered standard. However, cardiac disease continues to occur even at those cholesterol levels. Therefore, the standard must be lowered still further. The conclusion is that medicine still does not have the correct standard in place, so the level will be lowered until cardiac disease no longer occurs at that level." A very large number of individuals have now become "ill" because their levels, while once normal, are now seen as abnormal. They are seen as abnormal rather than researchers considering that the initial hypothesis about the relationship between cholesterol level and cardiac disease might be more complex than assumed or even false. The same scenario is occurring for high blood pressure and late-onset diabetes.

The field of addiction creates similar ideological classifications. Alcohol dependence has changed from true physical dependence being demonstrated to consideration of the number of legal problems the individual has accumulated. Smoking has become a universal disease generator, with advocacy studies indicating all human health problems are caused by smoking. The term "addiction" itself has been changed from a physiological description applied to the use of certain psychoactive drugs to a psychological description applied to any human behavior, even being in love.

There are a host of problem syndromic classifications: attention deficit hyperactivity disorder, Munchaussen's syndrome, Munchaussen-by-proxy syndrome, irritable bowel syndrome, restless leg syndrome, even Gay Related Immunodeficiency syndrome before scientists working on the molecular level found the actual disease caused by a lentivirus, HIV (20). Our geneticists tried to maintain the scientific standards perhaps because of the molecular biology level they generally work on. For that reason, they were reluctant without adequate scientific verification to decide whether a child was abnormally short because both parents were also short but healthy, or whether that child might have Turner's syndrome, which now has a karyotype test to verify it is a genetic disease and not a questionable syndrome. When they saw extremely wide-spaced eyes, hyperteliorism, they did not jump to a conclusion that this was a sign of a genetic disease. They first did the good science to determine that. They were more than aware with some of their patients of the FLK syndrome ("funny looking kid"), but they would not have considered such a classification to meet any of the standards of science, and would not have made it a part of the classification scheme of medicine.

Postmodern bioethics assumes that the failures to meet the scientific standard that are exhibited in medicine logically imply that scientific medicine is on the same, equal footing as alternative medicine. That argument is a logical nightmare, since there is no premise to justify what has been sneaked into the conclusion. All the failures logically imply is that some medicine is unscientific, not meeting the standards of scientific medicine. It cannot justify the failures of alternative medicine to meet those standards, nor can it justify changing or rejecting those standards. The example of Turner's syndrome is a good one. Once we had done the reduction to the molecular biology level, we then understood the process at the higher level of the child's body, and we could identify the mistake, the chromosomal abnormality, that results in the disease. That good scientific medicine gives us something alternative therapies are not capable of giving, understanding in a full systems sense, and with understanding, the possibility of corrective intervention that will reliably work.

There is also orthodoxy that is a human failure in the scientific system. The scientific method leaves no room for orthodoxy, but there are multiple examples of its human importation. It is one of the stubborn failures of the human condition that must be admitted, but also must be guarded against. There is no better safeguard than the scientific reduction itself, the rational requirements of the scientific method, and a modified correspondence theory of truth that, as WVO Quine said (21), allows one to finally deal with the unexplained "pings" of experience upon our hypotheses and even previously well-established theories. Science has constructured an epistemology that allows the persistent, amplifying pings that do not fit a current paradigm to be included, even at the expense of a major overhaul of the paradigm or the building of a fuller paradigm on the previous system of knowledge. Postmodernism is itself an orthodoxy, its tripod inviolable by the pings of experience, with no method to correct for new wonders and signals.

The postmodern politicizing of scientific medicine has unfortunately made it similar to alternative medicine in some areas, nutrition and diet being one. The ideology that has made lifestyle acts such as a proper diet the good acts of salvation has had its effect on scientific medicine as well. If one were to collect a list of foods that would give health and (implied) immortality or salvation, it would be a very interesting thing, and would make Woody Allen's movie, *Sleeper*, a very good bit of futurism. Butter, for example, was out along with olive oil, and hydrogenated margarine was in at the time of the movie. Now, hydrogenated fats are out, and butter and olive oil are reluctantly in, although an artificially produced butter-substitute is currently advocated as reducing cholesterol. Wine and other alcoholic beverages were out when the movie was made, but now, very reluctantly, a glass of wine is in for its cardiac benefits. Smoking is still out, but we shouldn't forget that nicotine has been shown in prospective studies to correlate with less Alzheimer's disease. Herbals may not be so benign as they appeared, and taking them in certain quantities may not be good for health. Garlic may or may not be a health additive for the diet. Both scientific medicine (not very well-done research) and alternative medicine have promoted the supermarket as the next pharmacy. There is even a darker side to this conversion of food into a religion. A mass population movement to reduce intake of food has resulted in a change in the population whose consequences are undetermined. This type of population experimentation has not been criticized by postmodern bioethics because, again, it suits the ideology. However, there should be strong ethical criticism of such mass experiments with so little scientific understanding to support them. These are major interventions in a large population of experimental subjects, and one would expect some comment on the ethics of research with human subjects. Postmodern bioethics had, from its inception, a strong component on the ethics of research and the priority of informed consent. The standards for judging such research to be ethical have always been extremely high, with information criteria that constitute a college course and with consent having a very high competence requirement and little tolerance for determinism of any sort. Despite this history of focus on the ethics of human subjects of research, bioethics has not criticized the social engineering and mass population experimentation in health and alternative medicine. Instead, there has been criticism in the *New England Journal of Medicine* of the ethics of AIDS/HIV researchers in Uganda and other African countries because researchers did not provide proteinase inhibitor cocktail treatment of the subjects and also followed the country's rule about not revealing the HIV status of the research subjects, even to their partners (22). Bioethics was willing to make an ethical analysis of such situations because, one could reasonably argue, the subjects were part of the ideology's favored victim class.

Alternative medicine, particularly naturopathy and homeopathy, engages in one of the ideological concepts of postmodernism, the politicizing and secular religion-creating of current environmentalism. Environmentalism is gathering all the functions and framing of a religion, and one part of the creed is the myth of the perfect balance of nature that exists before the intrusion of the human species. This new version of the Clockwork Universe runs counter to verified knowledge of systems dynamics found in general systems theory that accurately describes natural systems in a controlled state with oscillations kept within certain parameters by feedback mechanisms (23). Natural systems are not steady-state systems, and there is no perfectly-balanced and benign natural state within ecosystems, at whatever level one wishes to look. Furthermore, this oscillating and tentative balance has characterized natural systems far prior to the emergence of the human species. The scientific method gives the accurate description of the natural world, but alternative medicine supported by the ideology of postmodern bioethics clings to the myth. "Natural" things are good, notwithstanding there is no sensible definition of "natural" that would eliminate the acts of a species, Homo sapiens, and the products of those acts, technology, from the natural world. Natural things are good even when they are not, when the danger or risk is significant for their use. "Letting nature takes its course" is preferable when ideologically consistent with postmodernism, but without any argument on why acceptance or nonintervention in an ongoing status quo is ethically preferable. This conceptual mistake is basic to alternative medicine and raises ethical questions about its use, since such a methodology may easily produce unethical results with no self-corrective mechanism to lessen the risk.

The attraction of alternative medicine is its contrasting humanistic character as opposed to the cold, scientific, technological character of scientific medicine. Postmodern bioethics had already characterized scientific medicine as cold, intellectually-scientific, and frighteningly technological, so the ground had been prepared. In addition, as I hope I have shown in previous chapters, postmodern bioethics has also changed the definition of a physician and the clinical mode of practice. Bioethics redefined physicians as civil servants, as Engelhardt argued (24), or as college professors, as required by informed consent and truth-telling principles. The public was not pleased with the new character of physicians, now truly changed by bioethics into strangers at the bedside. Meanwhile, alternative medicine offered a personal and time-intensive relationship outside the medical system. Its practitioners were adept at emotional relationships and appearing to engage in a real relationship. Alternative medicine seemed more like the traditional physician relationship than the new creation of postmodern bioethics: the economic gatekeeper, the civil servant, the lecturer, the statistician, the salaried worker, the assister with suicide. The public preferred a human relationship, even if the science and good outcomes were lacking in alternative medicine. This may be one of the indications that the public has voted with its feet and pocketbooks against the new health care delivery systems, against postmodern bioethics, and against the lawyers, humanities professors, policy planners, and politicians who have taken control of the medical care system away from the medical experts, the physicians.

RETURN TO THE AGE OF HERBALS

There is an exercise we can do that is very timely. Since the 1940s, the human species has lived in an unusual period of time, when antibiotics gave our species some considerable advantage over the microbial world. We even had the hubris that allowed a health commissioner to state that the problem of infectious diseases had been solved for good, and they were no longer a threat to human health. We needed to remember the evolutionary fact that microbes exist on a much swifter time scale than humans, and that a crisis with medications has been developing since the discovery of penicillin. Before penicillin even left the laboratory, before we could conceivably blame antibiotic-resistance on physician prescribing habits or patients' demands, the first strain of penicillin-resistant microbe emerged in that laboratory that was developing penicillin. It makes good sense, because microbes have successfully survived on the planet from the beginning by competing with each other. Resistance genes likely evolved hundreds of millions of years ago in soil bacteria, because antibiotics were being produced by those bacteria themselves, as protection from other bacteria competing with them. Before human beings, before trilobites, microbes had to develop genetic engineering to stay in the race for existence. Biotechnology companies and medical researchers are newcomers to what microbes have been doing for eons: integrons, other transposons, R plasmids, conjugation and recombination, permiability barriers, channels, efflux pumps, analogs. All that genetic engineering was going on in the soil when life could not have conceived of a bioengineering pharmaceutical company, or a patient demanding antibiotics. We had only a brief window of protection, no matter what policy we chose.

That window is closing as everyone in scientific medicine knows. What if only alternative medicine from our history were left to oppose microbial diseases? Suppose we go back to my great-grandmother's farm and her time. What did she have to cope with and what could she do?

In 1866, there was a cholera epidemic in the U.S. and worldwide. Not much could be done about cholera except pray to the Fourteen Holy Helpers and hope home nursing would allow some to survive. Between 1868–75, New York had a smallpox epidemic, but there were vaccination techniques that gave us hope. In 1873, whooping cough went through the U.S., with no vaccine and again hope that supportive nursing would help some children survive. In 1878, a yellow fever epidemic hit the south, so my great-grandmother breathed a sigh of relief. Nothing could be done for yellow fever either. In 1889, influenza hit North and South America and Mary Kate could hope her family was partially immune as a result of previous infections. In 1894, polio appeared in Vermont, a bit too close for comfort. In 1897, dengue fever could be found in the U.S. Another epidemic of smallpox hit the U.S. in 1901–03, and even alternative American Indian medicine didn't do much for this. Between 1907–09, San Francisco had to deal with the bubonic plague, with nothing to help. By 1916, the U.S. had a polio epidemic and no medical response that could help, but neither could alternative medicine. On the heels of this epidemic, in 1917–19, the major pandemic of influenza may have killed 50 million people worldwide, and alternative medicine was again helpless, nor would praying to the Fourteen Holy Helpers guarantee a good outcome.

After the discovery of penicillin during World War II, a significant change occurred in the U.S., although not the developing world. There was a polio epidemic from 1942–53, before scientific medicine developed the vaccine that eliminated this threat. In 1957–58, influenza was still a problem, but the flu shots developed by scientific medicine also alleviated that epidemic but did not eliminate the threat, because in 1977–78, a pandemic hit the U.S. as well. From 1987–present, the AIDS pandemic affected certain risk groups in the U.S. Mary Kate did not live to see these last epidemics, but there was little she could have done to help with them. Her herbals might have eased the suffering, or even helped with getting a patient through some of the symptoms and increased the chance of survival. However, she would not have known what to use or in what dose, because polio and AIDS were new diseases beyond her herbals, and the viral nature of influenza was not known. She might have had some success with pneumonia as a complication, but not to the degree of technological support that can now be given, or the antibiotics that can still clear the infection, at least for a brief period as the window is closing. Her own medications from the fields probably included some antibiotics, though she would not have known the mechanism for their function nor the process they interrupted.

She and her church could have given hope and comfort. She could have cleared some infections with her unnamed antibiotics, her herbals and poultices. She could have nursed and cared for people. She and Dr. Nostradamus would have had something in common. He mixed his own formulas for medications, and had a good success rate caring for people with bubonic plague. He would go to areas where the plague had broken out. Both were good and decent healers, and I suspect they would have been very interested in our scientific medicine, eager to learn and use it, understanding they had provided the beginnings for its development.

REFERENCES

1. Mallory, J. P., & Mair, V. H. (2000). *The Tarim Mummies: Ancient China and the Mystry of the Earliest Peoples from the West*. London: Thames and Hudson.
2. Phillips, A. (2/8/2000). First century surgery. *The Medical Post*, *2*(28/2000), 53.
3. Burden, G. (3/7/2000). 'Doc' like an Egyptian: Ancient Egypt was a think-tank for medicine as a tome from the 13 century BC uncovers. *The Medical Post*, *36*(9).
4. Wood, M. (1997). In the Footsteps of Alexander. BBC: A Maya Vision Production. WWW.PBS.org.

5. Smith, F. G., & Stengel, C. J. (1988). *Anniversary Booklet: Fourteen Holy Helpers Church.* Fourteen Holy Helpers Parish, West Seneca, New York.
6. Hodges, D. (2000). Alternative Medicine Update: Rational Alternative Medicine. *The Medical Post, 5*(15/2000), 17–19.
7. Hodges, D. (2000). Is the burden of proof too easy? In: Alternative Medicine Update. *The Medical Post, 5*(16/2000), 17–19.
8. Rich, P. (1999). How Chinese herb led to kidney failure for Belgains. In: Alternative Medicine Update. *The Medical Post, 35*(12), 32.
9. Rich, P. (1999). Accupuncture: Most recognized but least used of the alt medicines. In: Alternative Medicine Update. *The Medical Post, 35*(12), 34.
10. Mallory, J. P., & Mair, V. H. (2000). *The Tarim Mummies: Ancient China and the Mystery of the Earliest Peoples from the West.* London: Thames and Hudson.
11. Brooke, J. L. (1994). *The Refiner's Fire: The Making of the Mormon Cosmology, 1644–1844.* New York: Cambridge University Press.
12. Rich, P. (1999). More nurses offering therapeutic touch but 'science seems to be nonexistent.' In: Alternative Medicine Update. *The Medical Post, 3*(16/1999).
13. Rich, P. (1999). Naturopathy tops list as vaguest alternative medicine, says expert. In: Alternative Medicine Update. *The Medical Post, 35*(12), 32.
14. Rich, P. (1999). Homeopathy cures so diluted they are useless, critics say. In: Alternative Medicine Update. *The Medical Post, 35*(12), 34.
15. Boyle, T. C. (1993). *The Road to Wellville.* Penguin Books.
16. Rich, P. (1999). Homeopathy cures so diluted they are uselss, critics say. In: Alternative Medicine Update. *The Medical Post, 35*(12), 34.
17. Gorovitz, S. (1984). Why you don't owe it to yourself to seek health. *British Journal of Medical Ethics, 10*(3), 143–146.
18. Sider, R. C., & Clements, C. D. (1984). Patients' ethical obligation for their health. *British Journal of Medical Ethics, 10*(3), 138–142.
19. Gabbard, G. (Ed.) (1989). *Exploitation in Professional Relationship.* Washington, D.C.: American Psychiatric Press.
20. Landau-Stanton, J., & Clements, C. D. (1993). *AIDS, Health and Mental Health: A Primary Sourcebook.* New York: Brunner/Mazel.
21. Quine, W. V. O. (1953). *From a Logical Point of View.* Cambridge, MA: Harvard University Press.
22. Angell, M. (1999). The ethics of clinical research in the Third World. *Science, 337*(12), 847.
23. Clements, C. D. (1995). Stasis: The unnatural value. In: R. Elliot, (Ed.), *Environmental Ethics, Oxford Readings in Philosophy* (pp. 215–225). New York: Oxford University Press.
24. Engelhardt, Jr., H. T. (1986). *Foundations of Bioethics.* New York: Oxford University Press.

10. BEHAVIORAL MYTH, PSYCHIATRIC ABUSE, AND SOCIAL MANIPULATION

THE POLITICAL USE/ABUSE OF PSCYHIATRY/ PSYCHOLOGY

When Barry Goldwater ran for President of the U.S., a public advertisement was taken out by psychiatrists who contended he was not mentally-healthy enough to be president. This public statement and political use of psychiatry was in direct violation of the American Psychiatric Assoication Code of Ethics, which states it is unethical for any psychiatrist to publicly make a diagnosis of an individual not seen in that psychiatrist's office or practice. The Code expected that a physician would have to examine a patient or look at a medical chart of someone who had examined the patient before coming to any diagnostic conclusion. The Code also expected that the obligation of confidentiality existed between a patient and psychiatrist unless extreme circumstances argued for setting it aside. If a psychiatrist were ethical, then, that physician would have a clinical relationship with the individual, in which confidentiality would almost always apply, and this would preclude any public statements in the political campaign. The political ad against Goldwater, which portrayed him as mentally unfit to be president, was unethical for two reasons: it would violate confidentiality if the psychiatrists had actually seen him or read the medical charts, with his permission, of someone who had seen him; and it violated the competence requirement that a psychiatrist would have evaluated the individual and therefore have scientific reasons for making a diagnosis. These are serious ethical violations for a professional, and there were many in the psychiatric profession who were uneasy with this use of the mantle of psychiatry as a political tool in a power struggle between two political parties.

Another startling example appeared in a psychiatric newsjournal, and concerned Saddam Hussein, shortly after the Persian Gulf War. It was written by a well-known psychiatrist, with excellent credentials, who insisted he was not violating the code of ethics because he was not really making a psychiatric diagnosis, having never treated or evaluated Hussein. His information was based on public descriptions and what others had told him. He then proceeded to make a very negative psychiatric diagnosis, despite claiming he was not. He informed the public that Hussein was suffering from Malignant Narcissistic Personality Disorder, one of the most negatively-value-laden, judgmental disease classifications in psychiatry. This classification is also one of the least scientific of the generally not very scientific personality disorder classifications. It involves remnants of psychoanalytic, self-psychology theory on which is imposed the unlawful and immoral behaviors of sociopathy. It is a nosology nightmare. That, however, did not stop a psychiatrist with a national reputation from demonizing Saddam Hussein, who had lost the Persian Gulf War with the U.S., with this psychiatric tool. The devious denial to the contrary, Hussein *had* been diagnosed without being seen and evaluated, and this diagnosis was not then kept confidential, but spread abroad in the marketplace (as the classic Hippocratic Oath would have worded it). It was an attempt to demonize an enemy through the unethical use of psychiatry, and there were no protests that it had been done. Instead, psychiatrists on national radio and television talked about Hussein's mental illness based on a videotape provided by news services of Hussein's meeting with foreigners detained in Iraq. The culture was busily demonizing Hussein, and this fit well with that goal.

When H. Ross Perot attempted to build a third party and threw confusion into the presidential race, the media began repeating a whispering campaign against him. It was implied that Perot was paranoid, and the implication then followed that he suffered from a serious psychiatric disease, paranoid personality disorder. I spoke with a reporter about the media use of the term "paranoid" as inappropriate, although most in the psychiatric profession did not speak up about this abuse of psychiatry. He felt "paranoid" was a general usage, not implying a psychiatric illness, but he did go to his standard dictionary to make sure it had more than a psychiatric meaning. He apologetically informed me it did not appear to have, and that paranoia was a psychiatric term. Since it was a psychiatric term, it carried all the baggage of a psychiatric diagnosis with it, and Perot was being labelled as mentally ill. The culture quickly picked up on this label, and his candidacy and party were dealt a serious blow. One cannot defend oneself from the charge of mental illness effectively. Once said, the damage is done. It is like gossip, and the old story about the woman told by her priest that her penance for gossiping and harming her neighbors was to throw open a feather pillow to the wind, and then try to collect each feather. This political strategy is pernicious, and to the extent that psychiatry cooperates with it, it is unethical and allows abuse.

The most recent abuse of psychiatry in this maner has occurred during the Bush/Gore election campaign, and it is a strange example. Gail Sheehy, whose books on developmental phases based on a self-selected population have somehow made her an expert in development and psychology, has, without being asked, diagnosed George W. Bush as having dyslexia (and can attention deficit hyperactivity disorder be far behind). Sheehy is not a clinician qualified to make this diagnosis, and she has never been asked to see Gov. Bush and evaluate him, but that has not stopped this writer on developmental phases from making the diagnosis and making it public. The same argument applies to her. It is unethical for a person who has not been asked to see an individual in their office or practice, or to read medical charts with permission, to make a diagnosis without seeing that individual. It is unethical to make a diagnosis public without the permission of the individual. Her statement about a dyslexia diagnosis was a political weapon used to attempt to personally destroy and discredit one of the candidates for the presidency, in preference to the other candidate. An unscientific and irresponsible claim was made that Bush's slips of the tongue indicated serious dyslexia and should disqualify anyone from the presidency. The claim was based on a series of minor misstatements made by the non-clinically-diagnosed candidate and printed without context. This claim about foreign relations was another non-expert lack of understanding of how foreign policy and world conferences are actually conducted, a mistake that anyone in history or political science could have easily cleared up if asked. But reality testing was not the methodology of this strategy. Rather, it was the deliberate abuse of psychiatry and medicine in the service of political power plays. This abuse also demeans and degrades those with an accurate diagnosis of dyslexia done by qualified clinicians, because it leads to considering them incapable of being in high political office. Since dyslexia also appears to be prevalent in my family in the form of transposition and ability to read from any physical perspective, and since my own reading and comprehension speed was once tested in the top 1% of the nation (with perhaps five people in that year slightly ahead of me), I have a personal interest in not promoting this degrading of any biological variance for political purposes. I could refer readers to Carpenter's "If Wittgenstein had been an Eskimo," but that would not undo the damage caused by this strategy that abuses psychiatry and general medicine. One cannot defend oneself against such attacks, even if, as in my case, one scored in the top 1% on the college entrance test. Having read this, the reader will still now begin to consider any awkwardness in my writing style as an indication I am incompetent because, perhaps, dyslexic. Because I, and some of my family, are also ambidextrous, perhaps neuroscience can provide a classification to demean that biological variance as well. Classic bioethics would have been very clear about this. All these individuals are acting in a highly unethical way, and classic bioethics would not have silently tolerated these abuses of scientific medicine.

The tool of mental illness is a particularly effective weapon against an individual, for which the individual is without a defense that is workable. There is no protection against the questioning of a person's mental status. Any response dignifies the charge to no avail. How does one prove one is mentally healthy? That question had bedevilled prosecutors in insanity defenses before the general change in the requirements of the insanity defense. Originally, since the state had the burden of proving the defendant guilty and the defendant was presumed innocent until such a burden was met, prosecutors had to show that the defendant was mentally healthy and not insane at the time of the act. This was a very heavy burden, and prosecutors often failed to meet it. After the Hinkley trial, the burden of proof was shifted to the defense. Now the defense attorney had to prove the defendant was insane at the time of the act and the prosecutor had no obligation to prove mental health. It is easier, of course, to prove in a court of law that a person suffers from such a severe mental disease or defect that at the time of the act it was not possible to form a rational intention, to know the difference between right and wrong, to know and appreciate the consequences of the act. Even then, most insanity defenses are not successful. The difficulties were compounded for the prosecutors under the previous system, who had the even harder task of proving the defendant was mentally healthy.

Once the accusation is made, there is always a suspicion, like the gossip feathers floating beyond the reach of the woman to take them back. Since we all exhibit natural and minor behavioral variations, we are all at risk of damage from such a use of psychiatry. Since psychiatry itself is not clear on the definition of mental health and what standards to use, and since it frequently errs on the side of seeing illness rather than health, there is already a pre-existing bias toward psychiatry seeing an individual as having some, even if minor, mental illness. There is no distinction between minor and major, except in a court of law, so minor variations from a culturally-approved norm can rapidly become part of the class of individuals with psychosis and extreme thought disorders. The slightly obsessive-compulsive physician could be seen as having a mild illness. The sad, chronically-depressed college professor may have a mild form. The teenager managing to show some sociopathic behavior by rebelling or the child being defiant could be diagnosed with a mild form. The intellectual who overrationalizes life might have a mild form. The self-sacrificing mother who is expert at generating guilt may have a mild illness. The list could go on as far as some psychiatrists have taken it, so that we could recognize ourselves in it and all of humanity. Although I would argue that all of humanity is not mentally ill and never has been (not even the self-mutilating mourning of male Huns who slashed their cheeks in order to cry with blood like men), and that our classification scheme, particularly with the personality disorders, is much too inclusive and culturally-relative, we still could all be publicly labelled. The use of psychiatry as political strategy and propaganda will harm us all.

Once it is permissiable to attack Hussein, Goldwater, Perot, or Bush using a psychiatric weapon, anyone else the centralized state or the total culture finds dangerous, inconvenient, or unorthodox can also be attacked. If this were being done to Russian "refusniks," we would be banishing Russian psychiatry from the international psychiatric organizations – as we did just a few short years ago. Then, postmodern bioethics was actively involved in criticizing the unethical use of psychiatry, perhaps because the refusniks were part of the victim class. The politics of using psychiatry against an enemy of Israel (another member of the oppressed class), or against political conservatives, seemed to fit the ideology of postmodern bioethics, and there was no active response, although this is a major bioethical problem in medicine.

The psychiatric tool is now an ideologically-acceptable one, given the psychoanalytic tripod of postmodernism and American culture's heavy involvement with psychotherapy both personally and in its institutions, including techniques of business management and mandated intervention in educational systems. My argument is that this tool is one of the worst forms of coercion, one of the most destructive of violent weapons. It kills the human spirit rather than the human body, but that makes its use even more ethically problematic than the typical forms of physical violence. Calling a person "sick" or "mentally ill" is a powerful thing. "Mentally ill" is even worse than "evil." An evil person has power and a perverse kind of dignity. The evil person has a purpose, a cause, a lust for living as he or she wishes, a commanding presence. The evil person can attract followers or grudging admiration. Some cultures considered Genghis Khan evil, but he was respected all the same. Some considered Nero the AntiChrist (1), but until current times, he was also respected. It was only when historians began considering he might have a mental illness and an ill relationship with his mother that Nero's person began to be diminished. Some considered John Brown evil, but again, it was only when he began to be portrayed as a mentally-ill fanatic that his stirring symbol during the Civil War was tarnished. The insane or mentally-ill person in our culture has no power and is degraded in our eyes. The madman invokes our disgust. The mentally-ill are separated from the rest of humanity, often in unfair stereotypes. There is considerable power to destroy a human being with the misuse of psychiatric labels and the long-distance diagnosis of mental disease, a power that extends across time. Consider mad King George. The king likely suffered from porphyria, and one of the symptoms of this disease can be irrational behavior on occasions. What is King George remembered for? He is remembered as being mad and having lost the American colonies. That is all a quick history leaves to a man.

In addition, the abuse of psychiatric labelling for propaganda gives mental diseases the very social stigma that professions have spent so many years trying to remove. It encourages a simplistic picture of mental illness that our culture has fostered.

With all these harmful consequences and unethical behaviors as described in classic bioethics, one should expect that postmodern bioethics would point out the ethical issues in such a political abuse of psychiatry. That has not occurred because the ideology of postmodernism allows any sacrifice for a political cause, and encourages the praxis of furthering neoleft social and political goals. Postmodernism's epistemology allows for making scientific and medical choices based on the ideological content of the conversation, even though it will not allow a content for the conversation of ethics. With its psychoanalytic tripod, it is comfortable with viewing the human world in terms of mental illness, and with the social activism of psychiatry, even suggesting social psychiatry as a praxis. It does not require a scientific basis for psychoanalysis or psychiatry, viewing both as narratives and the clinical treatment dialogue between patient and therapist. Botwinick argues that it rejects the theory in favor of the practice (2), which means that these narratives have no standards or concepts, are as empty of content as ethics, and as focused on process only. This is well-illustrated when clinical psychiatry/therapy meets forensic psychiatry. An expert witness must be concerned with scientific expertise, not cultural or political praxis/activism. The witness must make some attempt to be accurate, to have a standard of truth, to have considered facts other than the therapy narrative (such as school records, military, previous court records, family reports) (3). The legal system, although this is also changing, was set up to determine the facts. One clinical therapist complained during testimony that she was not a private investigator and didn't feel the need to determine what was true or false about her patient's statements. Unfortunately, the patient was suing a parent for child abuse in her past, and the court did need to determine if such abuse were a narrative or an event. The postmodern insistence that all is conversational process is incompatible with justice in the world, and leads to the creation of endless fantasy syndromes and potential harm. However, postmodern bioethics is not equipped to handle this ethical problem, only to make it worse.

THE NEW MYTHS ABOUT UNHEALTHY HUMAN BEHAVIOR

Early psychiatric intervention strategies, combined with the Health Belief Model's ethical assumption that it is always ethical to sacrifice individuals for population health, is a dangerous combination (4). For example, researchers are trying to identify children at greater risk for schizophrenia than the general population, based on markers of family history, social withdrawal, and problems with attention at school. Once identified, part of the group will be given the antipsychotic medications routinely used to treat diagnosed schizophrenic patients, to test the hypothesis that drug therapy before the first schizophrenic break may prevent or alleviate tht first episode of the illness. If there were clear scientific evidence for the markers being excellent predictors of later progression of the disease to clinical schizophrenia, this study would make some sense ethically and scientifically.

Schizophrenia is a severe organic brain disease. I recall interviewing a young man who had been a patient at a psychiatric center for almost ten years, who spoke and wrote poetically, who had been very intelligent, and who now could not fit into society. He still had the insight to understand his tragedy, which might have been the most tragic part of it. While talking about his history with the disease, he told me quietly that he could name the exact moment that his illness began. He gave me a minute, hour, date, and year when, he described to me, "there was a holocaust in my mind." I did not doubt that at the specific time he remembered so well something catastrophic had happened in his brain. He sat back, pleased he could understand what had happened. I have talked with other schizophrenics who have been hospitalized for most of their lives, one elderly woman who told her sister, the treatment team, and me that she felt as if she had been a Christian being tortured in a Roman coliseum. This disease can have terrible consequences, and the goal of this proposed intervention is very compassionate. The means, however, are ethically very suspect.

We do not have the capacity to make a scientifically-sound early diagnosis of schizophrenia, often not until the first major episode. We do not have a laboratory test, or a genetic test, nor do we know what gene or genes are involved and to what extent. We know there is a genetic component, from twin studies and family histories, so we know family history is a risk marker, but not a sufficiently accurate one for predicting that a child will go on to develop schizophrenia. If all we had were this marker, we would be making a major medical intervention without adequate knowledge that the child likely has this disease.

The other behavioral markers are so broad and vague that they could include all normal biological variances. "Withdrawl" is too broad. There are shy children whose passive response to the world around them could easily be normal. "Shyness" has not yet become a mental illness. There are children who are introspective and prefer a smaller number of social interactions that are more intense or intimate, preferring not to be social butterflies or team players. These children may seem odd to the team or group, but they find important social niches in the world and fill them well. Extreme intelligence is another biological variance which cannot fit easily into a group lifestyle because it is difficult to find others of similar interests and capabilities, although these children tend to blossom in a school environment for gifted children. They are variants, but are they mentally ill? There are creative and artistic children who see the world just a little differently and beautifully, and have their own aesthetic sense, whether Gothic or otherwise. Not all artists are schizophrenic, though most do not fit the American suburban model and historically most have not fit the, in their own term, "bourgeois" model. The effect of the school "in" crowd also needs to be considered. Such special groups can extrude individuals who do not conform to the cultural fads and fashions of the time, no matter how irrational they are. In this dynamic, which everyone who was a schoolchild will recognize, the withdrawal is forced by the social extrusion. It would not do to assume that the "in" crowd is the mentally-healthy group, while those extruded are the unhealthy. It could easily be the opposite way round. The conclusion is that social withdrawal is not a marker, at least a scientific marker, for anything specific to mental illness.

So far, if we considered the ethics of human research, we have not met the standards, at least in classic bioethics. Our prenatal detection program had rather high standards for doing population or couples testing, and I want to argue that such standards ought to be met except in situations where a good argument can be made for modifying them (5). One of the standards was *specificity* of the test (or marker). There are many ethical concerns about a test that generates too many false positives. It is a sensitive test, in that the searched-for disease is not likely to be missed. However, such tests create many harmful effects that outbalance the legitimate need for *sensitivity*. Individuals who actually do not have the disease are misidentified as having it, creating fear, disruption of ordinary activities, additional testing, and labelling. In most approved genetic testing, it was not too difficult to run follow-up tests with high specificity, then eliminating the worries and fears. There was still an ethical concern about the harm of even a brief misdiagnosis, and that concern had to be balanced with the availability of an easy follow-up test that was not too intrusive, in order to justify the ethical acceptance of a high percentage of false positives (6). For some of the clinicans, even a 20% false positive rate was unacceptable even with these safeguards.

In both genetics and in the testing for AIDS/HIV, the balance between false positives and false negatives was always kept in mind. For example, the ELISA is a sensitive but not specific test for HIV. It is designed to avoid as many false negatives as possible, since it is important for individual and public health that HIV positivity not be missed (7). The potential harm of false positives was balanced by required counseling and a very quick second test, the Western Blot Test, which was very specific for HIV. Both, however, have problems in Africa, where an even more specific test, the polymerase chain reaction test, PCR, would be preferable if it were not beyond the economic limits of most developing countries.

Social withdrawal as a marker or early test would have far too many false positives for ethical consideration. Even when combined with the genetic marker of family history, because that marker is also too imprecise, it would generate too many false positives to even be considered ethical. Now consider adding another proposed marker, problems with attention. This marker too is so undefined, so common in the general population in certain circumstances such as disinterest in the subject, that it fails its task. Furthermore, it is associated with a very danger-laden new behavioral disease, attention deficit hyperactivity disorder or its many changes of name. The questions of diagnosis of that new illness add another dimension to the problem.

There are no good lab tests for attention deficit hyperactivity disorder, no adequate clinical signs and symptoms, and disturbing indications that it is a culturally-relative disease creation, sex-skewed, pressured by educators rather than physicians or parents, and showing an abnormal increase in diagnosis that indicates poor science and possible manipulation for social control. It appears to be part of a growing trend to medicate very young children, before the possibility of a clinically-sound diagnosis can be made. In addition to Ritalin, we are now adding Prozac for pre-school children, and the study we have been discussing is proposing to also add the major antipsychotics used to treat schizophrenia. In a private high school with affluent families, only four of the teens in one class were not taking Ritalin. Of those four, two were taking Prozac, one had another psychiatric medication, and one student, one, was not taking any psychiatric medication. These are silent bioethics issues, not culturally promoted as assisted-suicide was, not part of an activist campaign with test cases, and yet affecting many more people. Postmodern bioethics does not appear to be able to handle such issues. The legal system, however, has responded. A class action suit is being brought against the manufacturer of Ritalin, against the American Psychiatric Association, and against the

American Psychological Association for colluding to promote the inappropriate increase in use of the drug, and the unscientific misdiagnosis of attention deficit hyperactivity disorder.

An early behavioral/psychiatric intervention program between schools and health care professionals can have many questionable results, in addition to the problem of accepting too many false positives. Early intervention itself is questionable even if the diagnosis could be confirmed by lab or genetic tests that are reliable. To make an early intervention, we should have to show it will profit the population to be diagnosed and treated, and that it will profit more than not being diagnosed and treated. What are the effects of diagnosing a "different" child as early schizophrenic? What are the effects of turning highly active children into sufferers of a psychiatric illness? What are the effects of defining a very young and sad child as having an illness rather than an unhappiness? The ordinary robustness of life, and ordinary sadness of life, the reality of the human condition that needs to be learned and built upon with maturity, all become reduced to a major affective disorder, an attention deficit hyperactivity disorder, an early sign (social withdrawal or difference) of a major psychotic disorder. All of this has a chilling effect on anyone who sees "outside the box" in the normal, creative way that many do. It could impose a subtle censorship on any mind that tries to understand reality outside the accepted cultural ideology, the accepted political system. Early behavioral intervention is every bit as dangerous, perhaps more, as it was claimed eugenics was in the early twentieth century. Postmodern bioethics, except very early in its history, has not analyzed and educated about this danger, has in many cases promoted such a health belief and such population control. Again, the answer is in the ideology of postmodernism, which favors such centralized programs of health and welfare, and which applies cultural pressure to reach a consensus based on political process, not on content or the freedom of the individual mind. There are no free minds in postmodernism, and there is a requirement for team programs toward social goals, the praxis. Those with variances who choose not to engage in praxis are at high risk, and bioethics will not advocate for them. They may be medicated into the team process, therapeutically-coerced into the ideological dialogue of the process, identified as having a mental illness requiring enforced help to join the process, or, if necessary, labelled as mentally ill and extruded from the process. Postmodern bioethics will not ethically criticize the new behavioral myths and interventions because they are part of its process and an outgrowth of its ideology. It wishes to use the conversations of psychotherapy for its praxis toward social goals, toward the total society.

This same process is occurring in pediatrics, where pediatricians are becoming, according to a local news item, "more than doctors," becoming counsellors and surrogate parents. The new role for pediatrics involves psychotherapy with adolescents, health education about proper parenting, and playing the role of parents to their young children; in other words, triangulating the parent-child relationship. Those involved in the training of pediatricians were pleased with this biopsychosocial model, but in terms of family systems, the traditional model places less disruption on the system, since the pediatrician then was not interested in assuming the responsibilities of parenthood in reality or playing at surrogate parenting in fantasy for an emotionally trusting and vulnerable child. The family system is too primary and vital to manipulate it through quasi-relationships, pseudo-parenting or surrogate parenting. These activities disrupt the balance of the family system.

Such active interventions can be extreme. In a child and adolescent psychiatry case, one mother had been so assertive in requesting treatment for her young daughter that, against the objections of her pediatrician, the child psychiatry consult suggested a diagnosis of Munchaussen-by-proxy. The child psychiatrist wished to call child protective social services, but the pediatrician opposed such a major intervention for a questionable diagnosis like Munchaussen-by-proxy. The ethical argument was grasped by the pediatrician, that a major intervention that could easily disintegrate the family could only be justified for very serious, well-verified reasons.

Unfortunately, new pediatric practices are willing to make such unjustified, major interventions. The triangulation of the family system is apparent in the words of one pediatrician: "Many times I'll say to the family, 'This child needs his own advocate.' I'll call the court and suggest they get an advocate." This pediatrician admitted pulling away from "pure medicine" into "social medicine," and thought this was a very good thing. She was not alone. The national Pediatric Review Committee's new guidelines strongly recommend that medical students receive formal training in behavioral pediatrics/ interventions, and the American Academy of Pediatrics focused on family social issues, including one year's emphasis on violence. The chairwoman of the committee saw the effort as bringing knowledge and love of children to the table, with pediatricians ideally suited to helping families. In the name of love of children, any major intervention was ethically acceptable because the intention was so good, even if the effects on the family system were very harmful.

This social goal for pediatric medicine, which was now social medicine instead of pure medicine, included: injury prevention by promoting the use of seatbelts and bike helmets; advocacy for children with career-pressured parents, absent fathers, violence in the home, etc.; direction on education and behavioral issues such as ADD; divorce; gun safety; home safety; date rape and sexually transmitted diseases; drug abuse; birth control, day care . . . The role of social medicine is endless, and that should be ethically daunting for bioethics. The ideological assumptions behind this social medicine are not disguised, and it can be seen that they are congruent with the ideology of postmodern bioethics. Because they are congruent, bioethics will not criticize this direction in medicine; rather, it will advocate for it, no matter what the risks and harms to families and freedom. There is the ideological assumption that such interventions help and strengthen families, instead of disintegrating them. However, the last thing we need is to substitute professionalized parenting and professionalized families for the actual family unit, even if it could be done. Pediatricians can only be actual surrogate parents if they are willing to take on all the obligations of a parent, to be a parent. Quasi-parenting is an illusion that should not be supplied to vulnerable children. It can only supply false promises of parenting, and to do that to a child is highly unethical within classic bioethics. A child's cartoon summarized this well: "There's nothing more cruel than a kindness falsely given."

Classic bioethics makes the argument that professional intervention needs to be carefully considered, especially intervention in the family system. It should only be done when a good outcome is specifically defined and reasonably expected, and when that outbalances any harmful consequences; when respect is recognized for the persons involved in the intervention, and when the methods used can themselves be ethically justified within a systems bioethics analysis. The risks for behavioral and social interventions are at least as high as risks for complex surgical procedures. These interventions need strong justifications and strong science behind them, and good intentions or ideological compatibility are insufficient for any argument. Pathologizing the family by making professional interventions may make the family weaker, as may supplanting the family temporarily with professionals. Creating child advocates divides the family into adversarial or power positions, destroying its needed bonds (8).

Expertise may be pseudo-expertise if not scientifically-based. There are multiple theories about child rearing and development, with very little outcome studies. Professional manipulations can be abused by political or social institutions for their own purposes as I have argued, and when those institutions have been politicized into a postmodern consensus, the risk to liberty is significant, not hysterical. The further medicine moves from pure medicine to social medicine, with the blessing of postmodern bioethics, the greater the risk of being coopted for behavioral and social control; and the further medicine moves from scientific medicine, with the active pressure of postmodern bioethics, the easier it is to coopt medicine for those ideological goals. Social medicine is a redefinition of medicine into a system that manipulates mass populations (without the ethics of research or the scientific design of research) in order to achieve the postmodern goals of centralized bureaucratic power. There is no ethical justification for such a redefinition of medicine except within the ideology of postmodern bioethics, a content that has surreptitiously been brought into its otherwise nihilistic philosophy, and a content that is narrowed to politics and power strategies. In the case of pediatric medicine, the medical tool can be used specifically to triangulate the family system, one of the few remaining systems that stand between the individual and the total society. It can be done in the name of vague social ideals such as advocacy for the children or home safety that have no scientific standing, can be done more easily if the scientific base of medicine is rejected, and can be done as an ongoing project to politicize the institution of medicine, a project that has been close to successful completion. This project has always been an explicit part of the change from classic bioethics to postmodern bioethics.

SOCIAL MANIPULATION AND THE SOCIETY OF GROUP THERAPY

Another aspect of the psychoanalytic leg of the postmodern tripod is the gradual intrusion of pschotherapy into ordinary activities of life. Since therapy has been opened up to practitioners other than physicians, a host of health care professionals or workers have become active in the lives of all of us. The management style of most American companies has incorporated therapy theories as part of the common managerial response, and also as special training sessions for management or employees. School systems begin therapy interventions with counsellors from the kindergarten level on. Religious institutions are heavily involved with therapy based on the secular, not religious mode. Police departments have their own therapists or contract with professionals. Law professors have argued that the legal system should become a "therapeutic jurisprudence," not a dispenser of justice and punishment (9). The average volunteer emergency medical technician and firefighter will also encounter mandated or expected counselling.

In a small rural community where everyone knew everyone, a local call came in for a pedestrian down. When the first responders arrived on the familiar street, and the address that one of the responders recognized as her substitute babysitter, a teenage girl was lying on the road. The girl's mother was standing in horror and the driver of a truck stopped nearby was in grief. He knew the family well. In the rural dark, he had struck the girl who was walking along the side of the road. The EMTs took over resuscitation from the fire chief who had run from the station nearby, and continued in the ambulance, but there was no response. The physician in the emergency room also tried to resuscitate the girl but could not. After trying for some time, he took the EMTs aside and said there was nothing else they could have done. Two of the cervical vertebrae were destroyed and the spinal cord was severed. No one could have brought her back. They left knowing they had done all they could, talking to each other on the ride back about the tragedy and the loss to the community. All the responders knew the girl, and her mother had been a member of the auxillary society.

Memories of the girl were already being collected in the community grief of small towns. The normal, real compassion and support, the naming of the sadness about losing one of their own, the gathering together, were already taking place in a natural way. But something else was also taking place. Emergency responders are a captive audience for trauma counsellors, and those counsellors were already at the fire hall when the EMTs pulled back in. All the fire department volunteers were there, summoned for counselling. It was required.

The EMTs found themselves sitting in a circle of chairs with counsellors who did not personally know anyone in the group. They listened to a therapist asking everyone to tell how they felt. The talk went around the circle, formal talk, impersonal talk, structured talk. One of the EMTs wasn't a talker, though. The counsellor insisted it was important to talk about the event and the feelings she had. The woman firmly said that if she had to talk about her feelings, she would talk to her partner as she often did, or she would talk to others in the community. She was not interested nor did she need to talk with professional counsellors, and most of the other responders said the same thing. Psychodynamic early intervention did not go well.

An ethical analysis would start with the freedom to refuse to participate or be called to such sessions. The EMTs were not given a choice, and that should be questioned. Is there any justification for mandating such counselling when those receiving it have not been asked for their consent, nor informed about the therapeutic intervention? My argument would be that such coercion needs to be justified in the same way that we justify involuntary commitment or other involuntary treatments, and not made a requirement for continuing employment or certification. Bioethics should not force therapy treatment unless there are strong reasons of the same quality as in other treatment choices. There are side effects of talk therapy that can be as dangerous as any medical procedure.

In addition to this ethical problem, there are other ethical concerns for classic or systems bioethics. There are problems with pathologizing helpers, with assuming that EMTs, police, firefighters, nurses, physicians, and others haven't the human skills and strengths to do what earlier generations have done, without therapeutic intervention. The normal human responses and coping mechanism operate in normal crises as well, and there are costs to pathologizing these expected, natural events that are so typical of the human condition.

There are problems with assuming "stress disorders," another syndromic classification open to abuse. Determining the level of malfunction that would qualify as a disease is difficult. There are also ethical questions about lack of screening in such trauma or crisis counselling for contraindication of psychotherapeutic interventions. It is neither scientific nor ethical to assume that intervention is not contraindicated and will be helpful. Such an assumption is not made for other medical interventions. There is also an assumption that it is curative or preventive to elicit feelings, to go over experiences with a counsellor, to express feelings in a formal, structured and impersonal way. Alternative means of coping are not given approval and variance is not allowed, which again requires that everyone act in a consensus.

Ethically, we should make a clear demarcation between medical and social problems, between abnormal variance and normal variance, between disease and life. Ego strength is a precious value that we all need to nurture and develop in the real world, the world apart from social-medical/psychiatric interventions. Being placed in the category of "the helped" without consent or need can be damaging to ego and not at all helpful in that maturation process. Postmodern bioethics does not have the vocabulary or the focus to even begin to consider this ethical issue.

For most of my career, I have been opposing another form of postmodern bioethics' social manipulation in the name of bioethics, the issue of sexual abuse and sexual harassment by physicians. From one of my first articles in a psychiatric journal (10) to expert testimony in a Montreal trial (11), I have argued that it is not unethical for competent, consenting adults to have a love affair, even in the environment of medical care and therapist's office. It has been a long argument, often upheld in the legal context, but generally disregarded in the codes of ethics of professional organizations, and in the majority of literature on the issue (12). A recent plea by a physician that his medical newsjournal support physicians in the latest round of task forces on sexual abuse of patients/clients by health care workers is a justified complaint about the incorrect ethical assumptions behind the sexual abuse issue. For that reason and with some fatigue but no change of mind, I looked at the Canadian

Health Professions Regulatory Advisory Council (HPRAC) report which looked at how professional colleges for 23 regulated professions dealt with cases of alleged sexual abuse. I also looked at another special task force, with its attendant public forums and its preconceived judgments, reporting to HPRAC (13).

In some ways, I feel like the Templar Grand Masters who rushed into breaches during battles. During the battle for Ascalon, a breach in the wall allowed Bernard of Tremelay to charge with forty Tempars into the fortifications and pay with his life and the lives of all who fought with him. Learning courage but not caution, the Templars repeated the charge into the breach at Mansurah years later, where they were cut down by street fighting tactics (14). The leaders of the medical profession, both in the U.S. and Canada, have not wanted to use such risky tactics, trying instead to coopt or cooperate with those charging the profession with unacceptable rates of patient sexual abuse. The medical profession has taken into its bosom the leaders of the attacks against it, hoping those attackers will see that the profession is earnest about being ethical and willing to work to prevent all unethical behavior by its members. That approach resembles more the Templars' surrender and safe conduct treaties with Saladdin, who reciprocated by beheading all the men who surrendered to him. There is something more worthwhile about Bernard of Tremelay and William of Sonnac charging through holes in the defense for what they thought was an ethical cause.

The profession of medicine is not the only institution which has compromised its integrity to a false ethical position which has a political power strategy at its core. The U.S. armed services have gone the same route, with ever-present task forces and special oversight committees. Stephanie Gutmann makes the same point for the military (15). She quotes a feminist who outlines the real battle: "cultural forces . . . at work within the military . . . dictate the harassment of women." The postmodern goal was to change the values and practices of the military using sexual harassment as the weapon, and it turned out that the main allies in this fight to politicize the military institution were the high-ranking military officers. Gutmann describes how the leaders of their institution tried to show they were attempting to change their culture by doing study after study, identifying the incidence of sexual harassment, and staffing those efforts with the very people who "most distrust military culture and who find sexual harassment around every corner."

I feel some responsibility beyond my expert testimony in abuse cases to oppose the current situation, because it was postmodern bioethics that was used as the tool to change the medical culture in many unethical ways, patient sexual abuse being only one of the social manipulations, and I was involved in bioethics, even if not in postmodernism. Sexual abuse became a wide problem, extending beyond a few psychoanalysts who loved their patients, when it was included in revised professional codes of ethics in a newly-defined way. This definition developed by postmodern bioethics stressed that a power imbalance innate in the medical culture made consent by a patient impossible, that a patient was so feeble and infantilized that she could never have an ethical love relationship with her physician. (The patient was usually female, and homosexual cases were rare, perhaps because homosexuals also belong to a specially-entitled victim class in postmodernism.) The new definition also broadened the meaning of sexual abuse and harassment beyond any hope of sense, making harassment anything the complainant felt it was. Gutmann shows what a powerful tool that expanded definition was in the military culture and the damage it has done to hope of an effective fighting system.

I was not surprised that the new HPRAC report tended to claim a terrible situation that has not improved and that requires more regulation and punishment, less niceties of legal rights, more structuring of women as weak victims rather than strong persons. The HPRAC report stressed that 100,000 Ontario adults say they suffered sexual abuse by a health professional. This number was obtained from a survey in which 1% claimed they had sexual contact or sexual touch, whatever those vague words mean and they could mean anything the perception of the claimant wishes them to mean. The HPRAC chairman thought this was a huge number, a "lot of people," but that too needs to be put into some rational context. I would argue that if only 1% of the population reported a robbery or other criminal activity, that figure would be a very positive indication that public safety was being maintained. The chairman of HPRAC was not happy because the objective was to stamp out sexual abuse completely, zero risk and zero tolerance. How a rational person could conceive that sexual behavior could be stamped out completely is puzzling. Human beings are sexual, another concept that postmodern bioethics recognizes only for multiple sex partners and recreational sex (free choices not to be restricted by intolerant religious institutions), and not for persons in a medical environment. Sex is also a complicated human behavior highly nuanced. How could any group of experts create as their objective the elimination of questionable sexual behavior to zero? Not even St. Augustine would have created such an unrealistic, inhuman objective. To achieve such zero risk (as with all goals of zero risk), and to use zero tolerance (as with all use of zero tolerance) for a vague definition such as sexual touching, contact or inappropriate remarks and behaviors, would require such unethical tactics that an ordinary person would and should be repulsed. The chairman complained that five years after the start, such behaviors have not been eradicated. That could easily have been predicted. Still, it did not deter the HPRAC from trying.

How did bioethics become so misconstrued? Some of the conceptual mistakes follow:

(1) The Freudian concept of transference did not initially include any major distinction between falling in love in other places and falling in love in a doctor's office (16).
(2) Transference did not mean that the patient could not make choices, lacked power, was a rag doll or puppet. It meant some choices were partly determined, often risky, and sometimes mistaken for the person making the choice.
(3) The culture of any institution determines individuals' actions and choices, but that does not mean we are non-responsible for those choices, nor that we are slaves or robots.
(4) The definition of a crime or unethical act cannot be in the mind of the complainant, but must have some objective reference in the real world, otherwise it is a dangerous, nonsense verbiage capable of great harm (17).
(5) A professional code of ethics is mistaken when it evaluates sex as prima facie wrong between any two people who are not legally incompentent.
(6) Women should not get special treatment (preference) unless they wish to be the oppressors or the victim-players-for-power. Neither is ethical.
(7) Professional leaders who support public manipulation on important issues are not behaving ethically toward the members of the profession or toward the public.
(8) Experts should not fail in their ethical obligation to be rational and scientific.
(9) It is sometimes necessary and ethical to confront rather than compromise.
(10) Current codes of ethics are neither rational nor scientific, and from a philosophic perspective, are very illogical ethical positions, based on a power-driven consensus.

Shortly after the HPRAC report, a Nova Scotia case illustrated the inherent weakness of this postmodern bioethics position on sexual abuse. Dr. William Hunter Blair, a general practitioner, had been treating Joanna Hyde with psychotherapy after the death of her mother. In the course of treatment, feelings of love developed between the two. Dr. Blair terminated with Ms. Hyde and then began a sexual relationship with her. One of his colleagues, not Ms. Hyde, reported him to the professional college. The case was investigated, although

Ms. Hyde repeatedly insisted to the college that she was happy with the current state of her life and did not wish to bring any charges, since she did not perceive a power imbalance, had competently consented, and was not harmed. At that point, she was known as Patient X, although her ethical requirement of confidentiality had been partly intruded upon by the colleague's charges. The college found Dr. Blair had violated the code of ethics against any form of sexual abuse at any time in the life of an ex-patient and suspended him for one year. He was also fined $4,000 in costs and required to have a psychiatric assessment at a sexual behaviors clinic. This last requirement may be the greatest risk to his human dignity. Because Dr. Blair was the only physician in the community, his suspension left the small community with no physician, but the pleas of community leaders not to take away their physician were not listened to. The college insisted on zero tolerance, with its Registrar repeating the ethical mistake embedded in the new code: There is no such things as a cooling off period. You can't just break off [a] doctor-patient relationship because you decide you want to have sex . . ."

Ms. Hyde then called a reporter and became a public figure, with her previous psychotherapy treatments a matter of public record. Her argument was that the new code of ethics damaged her self-esteem, her dignity, her human respect. In fact, her lawyer claims a human rights violation (18). She learned that this code of ethics would not allow her to leave the category of a therapy patient, even after termination, that she would always have the shadow of psychiatric illness over her life, that her normal ego strength and personal power had been taken from her by placing her in a special victim class where she could never have the capacity for consent. She had been infantilized and humiliated, and she knew it. The code insists that her falling in love is an indication of her illness, that she is not capable of normal love with a physician because she is acting out the transference and resistance. It makes her actions a mental illness not found in the DSMs and not verified by any scientific research. In fact, even Freud discounted this explanation. He admitted that the woman did really fall in love in a normal way, but that the physician must insist it was resistance on her part and tell a half-truth. That half-truth has been told in psychotherapy ever since. However, it is only since the 1970s and 1980s that this half-truth has become the basis for a major, rigid change in codes of ethics, argued for by postmodern bioethics and its feminist ideology component. The extension of this bioethics has spread the half-truth throughout social institutions, as psychotherapy models and concepts have spread throughout those institutions.

That this bioethics is a power strategy is implied by its empty terms that expand into more and more contexts, and by its zero tolerance for any claimed behavior. Those empowered by the strategy are not the women who claim sexual abuse. Those women must play the victim role, a role by definition without power and demeaning, and argue they have no capacity to make a decision about their behavior. Those empowered are the advocates of the claimants who can use the sexual abuse or harassment charge or threat to become dominant in the profession, to finally achieve the goal of 'oppressors' of the 'oppressors'.

The base of this postmodern bioethics is also very arbitrary about the application of zero tolerance. If the individual is a member of the ideological consensus, then the woman is seen as capable of giving consent (as Monica Lewinsky was), because the political praxis is the most important aspect of postmodern bioethics. If the individual is not a member of the consensus, then the woman (like Ms. Hyde) is made to be incapable of giving consent whether she argues that or not, because the political or ideological goal gained by finding there was sexual abuse is more important than the parties to the case, or than reason and ethics. This contradiction is a persistent and basic core of postmodern bioethics, and extends into other cultural institutions as well.

Using the psychotherapy model, postmodernism has extended itself across the culture under the positive face of the "helping professions." "Health" and "help" have such positive connotations that much can be superficially justified by appeal to them. However, a thorough ethical analysis would indicate that both concepts also have their dark side. In the case of health, too much concern with health, as some of the chapters have shown, can generate a risk aversion that is maladaptive and deprives human beings of the joy of their humanity. Risk taking is frequently a value, and too much prudence or focus on internal health states can be an evil. Health is also a changing status, and modern qualifications for "health" are extremely high, even in the health care system. Such high qualifications can do damage to human perceptions, to self-esteem. "Help" is also not always a positive thing. The helping relationship is difficult to successfully engage in, and has its own issues of dominance and lack of power. The helper is always in a one-up position in interactions with the person being helped, who is one-down., although this does not mean that the person being helped is therefore a slave who cannot make decisions (a simple look at patient non-compliance will cure us of that fallacy). Rather, this is a status and self-esteem issue again, with those being helped traditionally being recognized as having (and recognizing themselves that they had) an inferior status to those helping. If we neglect to take that into account, and again postmodern bioethics has no concept for that because it is neither a political issue nor a victim issue, the helping can produce great harm rather than good.

We are finally, in the area of social manipulation, talking about not only the abuse of psychiatry but the abuse of health care by postmodern bioethics in incremental fashion, in order to achieve ideological goals that are best described as a total society. For those who think such ethical concerns on my part are paranoid or an indication of mental illness, I would refer the reader to the movie, *Brazil,* or to an older movie, *Clockwork Orange*. If science fiction can show us the risks of the future, and I think it can to some extent, these two movies portray the risks of a health-care manipulation of populations far beyond what I have discussed.

REFERENCES

1. McGuinn, B. (1994). *Antichrist: Two Thousand Years of the Human Fascination with Evil.* San Francisco: Harper.
2. Botwinick, A. (1993). *Postmodernism and Democratic Theory.* Philadelphia: Temple University Press.
3. Clements, C. D., & Ciccone, J. R. (1984). Ethics and expert witnesses: The troubled role of psychiatrists in court. *Bulletin American Academy Psychiatry Law, 12*(2), 127–136.
4. Jeffery, R. W. (1989). Risk behaviors and health: Contrasting individual and population perspectives. *American Psychol., 44*, 1194–1202.
5. Clements, C. D. (1982). *Medical Genetics Casebook: A Clinical Introduction to Medical Ethics Systems Theory.* Clifton, NJ: Humana Press.
6. Clements, C. D. (1982). *Medical Genetics Casebook: A Clinical Introduction to Medical Ethics Systems Theory.* Clifton, NJ: Humana Press.
7. Landau-Stanton, J., & Clements, C. D. (1993). *AIDS, Health and Mental Health: A Primary Sourcebook.* New York: Brunner/Mazel.
8. Mirkin, M. P. (Ed.) (1990). *The Social and Political Contexts of Family Therapy.* Neeham Heights, MA: Allyn and Bacon.
9. Wesler, D., & Winick, B. (Eds) (1991). *Essays in Therapeutic Jurisprudence.*
10. Clements, C. D. (1987). The Transference: What's love got to do with it. *Psychiatric Annals, 17*(8), 556–560.
11. La Reine Plaignante c. Claude Saint-Laurent. *Cour du Quebec cambre criminelle et penale.* Le 17 juillet 1996.
12. Gabbard, G. (Ed.) (1989). *Exploitation in Professional Relationships.* Washington, D.C.: American Psychiatric Press.
13. Manzer, J. (2000). Conduct unbecoming; Two general patterns for MDs sexual abuse. *The Medical Post, 36*(17), 1–2.
14. Barber, M. (1998). *The New Knighthood: A History of the Order of the Temple.* England: Cambridge University Press.
15. Gutmann, S. (2000). *The Kinder, Gentler Military: Can America's Gender-Neutral Fighting Force Still Win Wars?* New York: Scribner.
16. Freud, S. (1946). On Transference-Love. In: *Collected Papers XXXII* (Translated by Riviere, J.), pp. 377–391. New York: Psychoanalytic Press.

17. Fuller, L. F. (1995). Eight ways to fail to make law, from the Morality of Law. In: J. Feiberg & H. Gross (Eds), *Philosophy of Law* (pp. 73–87). Belmont, CA: Wadsworth Publishing.
18. Walker, M. D. (2000). MDs former patient speaks out; defends their relationship. *The Medical Post, 36*(27), 5.

PART III

'BY THEIR FRUITS YE SHALL KNOW THEM'

11. ARE PATIENTS BETTER OFF THAN THEY WERE BEFORE THE SIXTIES: A SECOND OPINION AND ANALYSIS OF BIOETHICS

HAS THIS ANALYSIS BEEN TOO EXTREME OR POPULARIZED?

One possible criticism of the previous arguments from theory and cases is that the conclusion is extreme and does not follow from the premises. As a corollary, it might be objected that there were never intentions by the intellectuals who created the field of bioethics that were framed in such a conspiratorial manner. A second criticism might be that the cases referred to, the history given, and some of the bioethicists quoted were part of the popular culture and not the academic activities in bioethics.

Let me discuss the second criticism first. One of the key premises in my argument is that bioethics is part of the popular postmodern culture, and not a well-grounded academic discipline in the classic sense. Ethics is certainly a well-grounded academic discipline, and I have referred to many academic works in my analysis or in my articles which are referenced in my analysis. However, bioethics, my argument maintains, can be understood in terms of postmodern conversations in the advocacy culture and public relations, and cannot be sufficiently grasped in the traditional philosophic ethics of academic departments. Focusing there will miss the fundamental base in postmodernism, and the implications of that base for the practice and effects of bioethics.

In addition, bioethics itself has resisted being defined as a classic academic discipline, and the resistance of Ruth Macklin to my suggestion, in a conversation, that we use the framework of postmodernism to understand bioethics, was based on her perception that postmodernism was an academic term, of interest only to academics, and not applicable to bioethics in the clinical context. Bioethics has constructed itself as a popular or advocacy movement, with the creation of think tanks, the link with media, the heavy influence of the legal system, and the easily-identified political consensus that is part of its postmodern ideology. Wikler characterizes bioethics as an "activity" (a term we will recognize in the postmodern call for "praxis") or even a profession, as "opposed to a discipline or field of inquiry" (1). He uses phrases such as "a social campaign" and "advocacy." He recognizes that there is a somewhat dual character because of accumulating academic literature in bioethics and the activity of concept clarification to which some philosophy itself is narrowed, but this other aspect of bioethics can also be understood as public relations definitional strategy and the provision of subject matter for traditional philosophic study, rather than a discipline or field of study.

If there are two aspects of bioethics, the one involving activity and postmodern ideology is paramount by far, and this aspect tends to be lacking in self-examination or questioning of its core assumptions. Because it is essentially nihilistic about content ethics, it accepts the social conventions of the powerful of its time, with the charming hope of Ross that there were a multiplicity of somehow valid moral principles which were prima facie accepted unless a counter-argument were strong enough to dislodge them (2). These principles did not have to have coherence in any broader theory, and this, combined with the ethical nihilism in philosophy itself, gave easy credence to the principles of bioethics described by Beauchamp and Childress (3) and assumed without dissent from the very beginning of bioethics. These principles were vague enough and contradictory enough that the underlying postmodern goal of bioethics could be expressed through them. From that point, inquiry was basically closed in any real sense, and while academics might play with the margins, the political and social ideology of postmodernism was paramount. A quick review of Hastings' journals shows no theoretical movement from that point.

The other implication of the supremacy of the advocacy ideology in bioethics is that one will have to look to actions in the world of medicine, law and politics as the arena for an adequate understanding of what has occurred. Those actions are recorded in both professional journals and news stories of events happening and framed by bioethics. To record the actual activities or praxis of bioethics, one will have to look to reports of actions in the world of media, politics and law. The academy and its journals are not the field of bioethics, as expressed sometimes by bioethicists themselves. That entails a look at the popular culture, at case histories, at political movements and strategies, at all the public and popular venues in which advocacy shows itself. It turns out, at least for this ethical analysis of bioethics, that it is not what one says, but what one does that reveals the probable truth.

I am not alone in heavy referencing of media and culture, however. Most writing in bioethics also concentrates on such advocacy references, but not in a critical way, and not with the frame of postmodern theory that actually underlies that advocacy. The contribution the argument in this book makes is to identify correctly the real philosophic theory behind bioethics and its popular manifestations, and to make the connection between that theory and bioethics activities. The contention is that bioethics as a praxis cannot be correctly understood unless it is seen in its postmodernism context. There is a combination then, of theorizing and an historical, popular look at bioethics.

Let me address the corollary next, to clear the way for a summary of the validity of my conclusion. Is it necessary to have as one of the premises of the argument that there is a carefully-planned, fully aware, evil-intentioned conspiracy? I don't think this is at all required, though I am sure one of the counter-arguments will be that it is. There are cultural trends that it is assumed can be identified, and this assumption is not a new one in history. These trends are often the coalescing of philosophic or theological concepts, internal social dynamics, historical contingencies, and seminal events or literature. There is a social filtering that allows the acceptance of certain ideas and values, these find acceptance in the academic institutions and the cultural arts, and at some organizational point, political or power activities come into play. The human cultural system produces certain models, partly through human input, partly through the dynamics of the system. There is no need to postulate careful or even careless planning at this stage. The actors in this system naturally accept the model with a sufficient critical mass that the model then takes on its own determining power. Unless human beings are exceptionally good at critical thinking, this process usually occurs without much self-awareness or rational analysis and is unquestioned. Once the model is accepted, the goals automatically become good for those involved, and since there is not likely to be critical analysis, these good goals become enshrined. Along with Aristotle, I have generally assumed that the intentions of people are good unless proved otherwise, since there are few true sociopaths who only wish to do what they believe is evil. That is a rare condition, even among psychiatric patients who otherwise carry a diagnosis of sociopathic personality disorder. I think we have reasonably disposed of the concept of conspiracy theory, and the fallacious objection to any identification of trends as a mentally-ill production of conspiracy theorists.

One final argument in the refutation of this objection is that an entire discipline exists which involves the identification of trends and the probable projection of those trends into a specified future, Futurism. One example among many is the Club of Rome series of reports which uses a computer world model developed at Massachusetts Institute of Technology (4) to project multiple trends involving resources and industrial development into the twenty-first century. The sub-argument concludes, then, that there is no need to postulate a planned, evil-intentioned conspiracy as one of the required premises in this book's argument and conclusion about postmodern bioethics.

It is left, then, to look at the justification for the theoretical part of this book's argument. Is the conclusion too extreme to be warranted by the premises? One crucial premise is the definitional one of the "total society." This is a society which has narrowed its base to a political ideology of some sort, and which has coopted any systems within the society by politicizing them as well. There is then no intellectual or empirical or even traditional base from which to critically reject any of the goals or methods of this total society. By asserting that all is process/politics, any content within systems other than the political system becomes invalid and irrelevant. A sub-argument could be made that such a total society will be static and enforce a consensus based on the assumptions of those who have successfully used the political process to gain power. This society is different than a tyranny based on military force or the charisma of a leader, because it has enveloped all the systems within it to form its consensus and to maintain it. It is more than authoritarian or dictatorial; it is total in its methods and its spheres.

A further premise is that such a trend to a total society can be idenitifed early in its origin. A corollary of that is that it is important to make that identification, because this form of power organization is incompatible with human nature over the long-term and unethical. Such total political/ideological process runs counter to essential aspects of human nature, as slavery, although widely practiced in history, runs counter to essential aspects of human nature. (Human beings do not make good slaves as they must be kept under close control often with physical force, tend to revolt against their bondage because they innately do not like it, and always represent a potential danger to the slave-holders.) One of the important definitions of a human being is a free/reasoning/learning mind, and while this is an ideal that is generaly not met, it is our evolutionary strategy and works well enough that it is so far an adaptive strategy. That may coldly define it in terms of a systems bioethics, but on some levels of the system, it is even more than that. It is held as a value even greater than life and is considered the ideal accumulation of maturity. Since this value is seriously jeopardized by an ideology that allows the development of a total society, we can make a sub-conclusion that an identified ideology which leads to this would be an unethical, unwarranted ideology.

The next premise is that postmodernism represents such a trend, such an ideology, and would lead to a total society. Postmodernists would deny this, of course, because ironically, they see their ideology as leading toward the freedom of man, the full expression of humanity, the social good or welfare that aims for perfection and shares some aspects with utopian thinking. They see the requirement for praxis as the requirement for the good, although they illicitly import a neomarxist political view into the supposedly content-less praxis. In this way, they avoid the postmodern problem of the psychoanalyst without any content theory who consistently engages in conversation or dialogue with his patient as therapy, the process of praxis having to be content-less or, most concerning, based on power and coercion. They can also claim that importing content, which by their definition cannot be valid or verified, can be done in any conversation if one wished, since there are no logical or rational standards for such conversations. We add here another premise, that given postmodern epistemology, the only standard for "truth" or "good" *is* power or coercion, what a process, even a consensus process, is actually about, and only what it is about. The extreme skepticism or agnosticism (or nihilism) that Botwinick attributed to postmodernism is accurate (5). Any importation of ideological content, then, is done by power or force (negotiating a consensus is force as well as coming armed to the negotiating table) and is suspect. My argument focuses on the process of postmodernism as well, because it has denied the possibility of any valid or verified content, and looks at the mechanisms created within the social system to spread the postmodernist ideology, less importantly the ideology itself, although that is also relevant. Another premise follows, that the ideology illicitly imported into this basic nihilism of power-process is a neomarxist ideology which supports such a political reduction of systems and culture to political strategies or means, and has as its end the forced consensus of centralized authority and thought control. The globalization that postmodernism advocates is a centralized authority beyond historical precedent. A reading of the content consensus found in neomarxism/postmodernism illustrates that without the safeguards of reason or science, only certain concepts or positions are acceptable and others will be eliminated by the consensus, through the use of power dynamics.

The argument need not insist on a premise that requires absolute consensus or a monolithic program toward a total society that is also monolithic. No content program will be such a monolith. Even the Christian and Islamic frameworks in history did not achieve such a monolithic character, but contained sects within them. In a content program, there will be minor variations and interpretations, but they will be variations on a theme. If the variations become too dissimilar, they will be extruded from the content program, as the Christian and Islamic frameworks were extruded from the Jewish framework, or the Church of Latter Day Saints is frequently extruded by those within the consensus of Christianity. The important point about the postmodern consensus is that the process consensus is close to being monolithic, and that the content consensus is driven by the process consensus. That nearly monolithic process consensus has as its goal a remaking of all social institutions or systems into the process consensus, which will then lead to a total society that may have minor variations on a content theme, but will be monolithic in terms of its power process. The power process itself, and its refutation of any content base that might oppose it defines the total society of postmodernism.

For that reason, another premise is maintained, that we can understand the postmodern process by looking at examples of the working of the process in social systems, and that a prime example of its reducing a social system to political process is current bioethics. If we can show that postmodern bioethics has redefined, reconceptualized, reduced bioethics to a political ideology and process, then we have an example of how this process works in a society and what its implications may be, based on the results within that health care system. Part II of this book was an attempt to argue this premise and its implications. Using multiple specialties and multiple case examples, it attempted to show how the change in bioethics resulted in certain predictable changes in the medical care system, and certain predictable new concepts and conclusions for bioethics. It was an addition to the historical argument for the premise in Part I, and was a philosophic argument that identified the actual processes in thinking and enforcing that characterize postmodern bioethics. It also demonstrated the accepted consensus in bioethics and was a partial explanation for the inflexibility and exclusiveness of that consensus. If the empirical and theoretical arguments in Part II are sustained, and I think they are, then we are looking at a power process in health care, and by extension, in all social systems, that is producing an ideological total society. This does not mean that such a trend can be projected indefinitely into the future. That is a primary mistake of using projected trends. What it does mean is that the current trend is a dangerous one for human values and a dangerous one for the medical care system. There may be other counter-trends which will prevent the successful formation of a total society, but if they are present, they have not built up sufficient critical mass to influence the postmodern process and its continuation.

I do not believe my conclusion from these premises is unwarranted by them, or too "extreme." It is an outlier conclusion of an outlier argument, and in the postmodern process, such outliers are extruded. They are extruded not because they are invalid or unverified, but because they deviate from the consensus achieved through power dynamics. They are labelled "extreme" as a way of extruding them, as is most critical thinking, because in a power process critical thinking has little power. There is, however, one further premise I wish to consider for the argument: Are patients better off than they were before the 1960s and 1970s discontinuity in bioethics?

THE PRAGMATIC CRITERION: ARE PATIENTS BETTER OFF THAN THEY WERE BEFORE POSTMODERN BIOETHICS?

The answer to this question is a pragmatic argument that depends on concepts of workability in the human transaction with natural experience, as Dewey argued before Rorty modified pragmatism into postmodernism by defining workability as the activity of the transaction alone (6). It is easy to ask if a conclusion or hypothesis works. It is not so easy to define what we mean by working. The phrase "better off" is just another way of asking about the workability in ethical terms of postmodern bioethics.

Unfortunately, John Dewey wanted to eliminate consideration of metaphysics from his philosophic position, but that elimination has predictably led to a foundationless definition of "pragmatic" or "workable." We cannot really define pragmatic unless we return to a metaphysical position, otherwise we are left with process and no content and have slipped into the sound and fury of postmodernism. Our transaction with natural experience allows us, as Marvin Farber argued, to postule the concept of an external reality (7). The concept is necessary in order to explain the unexpected, the "pings" intruding on our present framework of theoretical explanations of the cosmos. We need not postulate that we can know the thing-in-itself. Realism does not depend on such a mistake, nor does the Enlightenment. We only need the highly-probable construct that there is something external to human experiencing, hypothesizing, wishing that is recalcitrant and expanding as our instruments for perceiving expand. We only need to consider that when we transact or ask natural experience questions, we get answers we neither wished nor anticipated, answers that sometime require the restructuring of our dearest human constructs. That would not happen if reality were merely a social construct, or if one narrative were on an equal footing with all other narratives. Sometimes, in fact, the answer is painful, and we do what any good realist or naturalist would do, we go back to the drawing board. The transaction process with the external reality postulate has a product: valid or verified content. The pragmatic criterion actually asks whether the answer we received from this transaction gave us a painful indication that something was incorrect with our hypotheses or affirmed that those hypotheses appeared to safely lead to new understandings and interventions in the cosmos. It is, to some extent and in philosophic terminology, the learning program again, the evolutionary strategy of being human. Patients, then, are better off if bioethics enhances their humanity, allows them to be fully human, supports their learning program and coping with natural experience, and affirms their species' interests ("it's a wise species that knows what its self-interest is").

What has been the record of postmodern bioethics?

From its inception, patients have been given the right to die or to refuse treatment; or to consent to be part of clinical scientific research; or to kill an entity in the uterus, in a coma, or in a vegetative state without asking the ethical status of that entity. They have not been given the right to live, to request and receive treatment, to refuse and be protected from mass population research and manipulation, or to have their status and worth as members of an ethical community affirmed. The actual result is that economic pressure and rationing have accelerated in the health care system, indirectly coercing patients into early deaths, less medical care, and formal procedures that do not protect research subjects adequately. Ill patients are isolated, with less power to determine their destiny than they once had, made to feel a burden on the health care system, allowed to die for utility/futiliy reasons, deprived of personal or physician advocates, and when discharged from the hosptial, subject without consent to mass population experiments for vague social goals of "health" as defined by the state and the medical arm of the state. Patients have not been given their full worth as members of an ethical community, but have had to prove their utility to the state or their sufficient score on a Quality of Life scale.

Bioethics has worked to restrict assisted reproductive technology and reproductive medicine, refusing couples the right to make those decisions. It has limited research and development in medicine in general, attempting to censor and regulate bioscience technologies. It has denied the human value of reproduction and the worth of human generation.

As a result of bioethics' support for downsizing the health care system and experiments in new delivery systems, hospitals have often failed at infection control standards and reporting infectious diseases. The Center for Disease Control added "and Prevention" to its name, and then focused on social goals such as violence "epidemics" and guns and obesity "epidemics." The actual containment of infectious disease outbreaks was ineffective: too little, too late. The managerial focus was on paper planning, coordinating planning, public relations brochures, and websites. There was less use of traditional public health control measures for very serious infectious diseases, including AIDS/ HIV, tuberculosis, West Nile virus, and pneumonic plague.

Causing a patient's death by passive or active means was legally and clinically routinized, and physicians became advocates for terminating patients, not all of whom would be naturally terminal or even suffering from a disease rather than physical condition such as a developmental handicap. Unhappiness became a standard for direct termination or assisted suicide in the advocacy of bioethics. Health care systems began pressuring (because of legal mandate) patients to look over a standardized living will form advocated by bioethics proponents of all forms of euthanasia, upon hospital admission for any reasons, even delivery. Families were pressured by staff to consider withdrawing or withholding treatment for incompetent patients who had not written an advanced directive. Patients could be denied life-saving treatment even when requested because of staff perceptions of medical futility.

Women, including underaged girls, were given the right to abort through to full-term, and no state restrictions were allowed. Membership in the human and ethical community was removed as a bar to harmful medical procedures. The ethical basis for human worth and respect was removed from discussion as irrelevant. Patients could be called "its."

The Health Belief Model allowed lifestyles to be coerced and manipulated by government, and populations to be indoctrinated. Social goals overruled personal needs and interests. No informing or consenting was required for population experimentation that was intrusive in important areas of life, such as diet, risk-taking choices, parenting, transportation, and education. New health delivery systems put a managerial class in control of the health care system, with various techniques for limiting access and utilization by patients. Bioethics supported rationing and management-heavy systems, as well as bureaucratic levels to operate regulations. Patients were moved as units throughout the complicated system, where utilities of scale overrode compassion and quality of care. Tertiary health care systems moved responsibility for certain patients to homes, nursing facilities, or outpatient care even when inappropriate. The new delivery systems continue to escalate costs and downgrade quality of care. Health became the patient's responsibility and illness the patient's failure or mental malfunction. Self-inflicted diseases were stigmatized.

Science was removed from the health care system as much as possible, while alternative medicine was reinstituted in both hospital and society. No quality control was advocated for alternative medicine. Scientific expertise was no longer a base for criticizing or opposing social goals, economic experimentation with delivery systems, or health fads.

Medical expertise was abused for political purposes, with abuse of psychiatric diagnoses being the most common and most likely to impact on anyone. The health care system was in place to become an arm of the political state. Social medicine began replacing scientific medicine, with implications for invasive control of individuals and families. The cultural is becoming psychotherapized but in a non-medical model with little science supporting the counselling, and a great deal of population control likely.

Postmodern bioethicists will object that these changes and consequences are not the result of their "activity." I hope the bulk of this book demonstrates that such an objection will not work. The ideology of postmodern bioethics is precisely the cause of these changes and consequences, and there should be accountability for it. Ideas and concepts do have effects in the real world, as much as medical surgical interventions, and in bioethics, there ought to be the same level of accountability that the previous decision makers in the medical care system had. Unfortunately, there is no professional or legal accountability for bioethics, the catalyst and apologist for the changes not only in medical ethics, but in the medical care system, that placed lawyers, humanities professors, theologians, and politicians in the role of decision makers in the health care system (8).

THE SECOND OPINION

A number of years ago, Dr. Roger Sider and I wrote an article, "The new medical ethics: A second opinion" (9). I had still not realized the actual character of the new bioethics and the importance of the postmodern framework in understanding its activities. We did have concerns, and realized a second opinion was ethically required, and we had an understanding that there was a difference in this ethics, that it was in some way without content. At that time, we suspected its content-less character was the result of its focus on individual civil/legal rights and autonomy. We knew there was an individual subjectivism underlying bioethics, but we were too concrete in seeing it as the exclusion of the values of the family, physician, significant others, and the system of generalized human values itself. We were beginning to see that this new bioethics was a political and legal model, and that it was incapable of dealing with normative medical practice, but we had not yet seen the conceptual framework that provided this ideological approach. We were still describing the medical care system with its scientific and clinical values, and not the emerging health care system that had neither. We did realize that this new bioethics was a threat to the fundamental value system of the medical profession, and that a demoralization and deprofessionalization of medicine was imminent.

This editorial in the *Archives of Internal Medicine* was a first attempt at a real understanding of what was happening in the name of bioethics, and it was inadequate to the task of correcting a major mistake in ethical history. Out of trust in the academic tradition, we failed to see the power strategy and our article did not help to hold the line. Nor could we have prevailed at the time, because we had missed the postmodern process. There were keys to the analysis we were missing. Our second opinion did not help.

A bit later, the importance of the history of medical ethics provided one of the keys. A timeline of medical ethics from the shamanic tradition onward was constructed in multiple formats, one on a roll of paper that could be opened up, as a lawyer did, across a conference table. There was the discontinuity and there was the key, evident in the almost battlemap strategy of a timeline. Three more articles followed from this timeline key, but there was still something missing in the first one (10). By now, my amateur paleontologist son had suggested the word "discontinuity" to describe the break in the bioethics tradition. It was my English literature daughter who inadvertently provided the second key. She mentioned postmodernism in literature, and that key clicked the lock. My argument is that we cannot explain the current bioethics, we cannot provide a reasoned philosophic and ethical second opinion, unless we appreciate that current bioethics is one of the activist forms of post-modernism.

It should not have taken me so long to realize that. Engelhardt was quite open about this philosophic grounding of bioethics (11). His book is an exposition of postmodernist theory applied in medicine, if it is read with that understanding. I had also done seminars for our physics department on ethics and science, and had begun to outline the postmodernist stance in relation to the sciences. When the plutonium experiment case was discussed, the postmodern assumptions about medicine and ethics were critiqued. The difference was that most bioethics discussions of postmodernism and its praxis are done with a positive evaluation, since done by postmodernists. The social activism is implicitly approved by Wikler (12) and Rothman (13), and a full, critical analysis is missing. One has to step outside the postmodernist theory to adequately critique it, because within the theory, illogic and unethical sacrifice toward good-intentioned social goals are accepted without question. One has to get outside the therapy office into the postulated reality of the natural experience. In a sense, one has to be humanly-healthy. In other words, the human learning program in its transaction with natural experience has to be operating, not malfunctioning.

That full and critical second opinion would have to conclude that patients are, in general, worse off now than they were before the changes of the 1960s and 1970s. From a classic humanistic ethics position which operates from the learning program that genetically defines the species, the following points can be made.

Killing a member of one's own species is an awesome act with potentially damaging consequences to all. Although it can sometimes be justified, it requires strong justification. It is the possible option in the clash of two conflicting primary human values: biological recognition of a member of one's species and attendant sense of sameness, as Lewis Thomas demonstrated for less-organized life-forms (14) conflicting with territorial and protective biological interests. When we refuse to acknowledge or when we attempt to make irrelevant that strong recognition of membership, we eliminate the ethical base for human worth, respect and freedom. If that is eliminated, the health care system moves toward discarding persons for political or economic goals which are unjustified. This places all persons at risk in the health care system, a system they entered with the hope of help. They will either be eliminated through sophisticated manipulation or through the process of legal rights; be in fear that such a thing could easily occur if they have no utility; feel unworthy and a burden on society; or recognize that they and health care workers are frequently adversaries. The same lack of trust that was common in the shamanic tradition will recur in health care systems, and the rippling effect of that legitimate distrust will make a necessary need for medical care also a high-risk choice. No one, not even those who do wish to die, is better off under such a system.

Another basic human value, reproducing the genetic program into the next generation, is also devalued in postmodern bioethics. The devaluing of an individual now extends to the devaluing of the human species, which is often described as overpopulating beyond the carrying capacity of the land, upsetting an otherwise benign balance in the ecosystem, and being like a cancerous growth on the planet. Assisted reproductive technologies are seen as unethical and a waste of resources. It will be more difficult for individuals to access ART, more difficult for funding of new research without overregulation, and more difficult for individuals to feel content with their legitimate choices. This death wish for the human species has been present before, in religious communities like the Cathars, but its practical effect of making ART more difficult leaves patients in worse shape than before the new bioethics.

Universal precautions, infection control, and control of outbreaks through traditional public health measures have been replaced by management-on-paper and limitations resulting from the assumptions of postmodern bioethics. The biological reality that an ecosystem is a balance of power among many levels of life-forms, and that the microbial level also has a successful evolutionary program, is made irrelevant to cost-cutting in the health care system and assigning responsibility to the individual to protect himself. This has occurred just as the atypical window of relief in this struggle provided by antibiotics is closing and new technologies and public health measures are required. The single individual does not have adequate means of protecting himself or herself from iatrogenic or infectious diseases, and it is an illustion to place responsibility on the person. In an economic first-order decision about cost and accepted lives lost from infection, the individual is always the sacrifice, and the new bioethics has justified that. The lives and health of patients are now at more risk than they were under classic bioethics, and these again are important natural values.

Similar values of personal comfort, hope and humane relationships are at high-risk in current health care delivery systems, systems brought into being through the support and suggestion of current bioethics. The patient is now thoroughly dehumanized and disempowered in such systems, becoming units that are part of large packages controlled by businesses, governments, or even health care insurance organizations, and negotiated back and forth among the real powers. The patient is not the customer, not even the consumer who decides what to purchase. Economies of scale and complexities of systems move this unit around for the internal efficiency of the system, not for quality of care. The value of taking care of a sick individual occurs even among less-complex life-forms such as wolf packs or foxes. It is another primary value that is lost in postmodern bioethics. With it, a social species loses important bonds that hold the social group together, and the loss of altrusim (or enlightened self-interest, which is basically the same) requires a human being who is sociopathic. There is, then, not only the concrete pain and personal energy expense demanded of "units" in the health care system, not only the lack of control and power, but the likely loss of human identity and human caring for others. The system will structure for sociopathy, because a sociopath views others as objects or furniture in his or her world, to be used as pleased or needed. We will all be worse off in postmodern bioethics.

The learning program must place great emphasis on correcting mistakes and supplying a pragmatic view of reality. With the relegating of science and its methodology for doing that very thing to an equal or even lower status than other stories, the health care system has no safeguard against harmful fads and fancies. The patient cannot be protected from harmful illusions and fantasies, even from psychoses, in such a bioethical system, because there are no longer any standards from the learning program to so protect him or her. Nor can the patient protect him or herself within the system, unless a professional within that system, even if the patient's learning program is working well. Patients are more likely, then, to be harmed than they previously were.

Social manipulation and indoctrination, now more common in the system, violate another basic human value, one that is a requisite for a successful learning program: human freedom. The biological nature of human beings programs free thought as a mechanism for learning, and free thought requires a degree of human liberty to operate. This is not a political freedom, which could be merely the access to voting. This is too narrow a freedom and will not alone safeguard liberty. Political rights or freedoms, given by the state, can be easily taken back by the state, or reinterpreted to mean what the state desires. A firmer ground is needed for human freedom, and it is found in the classic bioethics. It is not found in postmodern bioethics, which is easily inconsistent about rights and which is willing to sacrifice old ideas of freedom to the social goals of ideology. Large percentages of the population, under the new bioethics, are subject to social experimentation without their consent. Mass populations are subject to health interventions that are scientifically questionable. Populations are structured to consensus. This move toward a total society encompasses the health care system, and patients may again find themselves in an adversarial and authoritarian relationship with health care professionals who, in the social arena, now have more control over patients' lifestyles than was had under the classic bioethics. Patients cannot be better off if less free.

Finally, the professional value of competent, quality care was eroded, and the physician was no longer comfortable as a patient advocate. The physician has been redefined, and not for the better. An objective look at the history of medicine shows a healer with human failings as we all have, supported by a classic tradition of securing an objective good outcome for the patient that could be reasonably and philosophically determined, and held within an ethical community to such human values. Classic bioethics was content-ethics, and it is an incorrect assumption to argue that such content was invalid. It is only invalid within the current tradition of postmodernism and the erosion of philosophy that has unnecessarily occured in the twentieth century. Holt describes neutrally the result of the analytic philosophy of scientific propositions or ordinary language, the first coming out of Cambridge, the other out of Oxford (15). He discusses Ernest Gellner's asking what use is philosophy if its only function is "to relieve self-induced feelings of confusion," what more sterile an enterprise could there be than one that aimed to, in Wittgenstien's words, "leave the world as it is?" That need not have been. Marvin Farber once described this approach to philosophy as sharpening one's analytic knife until there was nothing left. Unfortunately, Farber's prediction came true, though it need not have. The legal play with words practised by bad lawyers began to fade and what Holt describes as the more technical sort of American philosophy became paramount. That philosophy was heavily political, with the American suburban, liberal ideology, as John Rawls' *Theory of Justice* illustrates (16), and it became more political. Postmodern philosophy then became descriptive of this philosophy, and we have come to where we are in ethics and bioethics.

Where we are was captured in an article in *The New York Times* in 1998. The report was on the World Philosophy Congress, the last of the twentieth century. The panel included Quine, Strawson, Davidson, Greene, Apel, and Nasr; and the question to them was: "What have we learned from philosophy in the 20th century?" Strawson focused on the ambiguity of the word "we." He did complain that too many believed only the last two decades of philosophy were worth attention. He suggested there were paradoxes to savor.

Davidson decided to talk about how very American philosophy had become, how "Harvard" but that e-mail might change this. Quine did not answer the question. Greene also played with the word "we" and well as "learned," and then asked why mathematical logic, why Descartes, why Heidegger. She did decide to evaluate Heidegger as evil, but then said ethics was "just minding everyone's business," which might mean that Heidegger should get an ethical pass. Apel was sorry that reflection on other minds had become nonsense under the analysis of linguistic philosophy and had been replaced by infinite mathematical series: "The only philosophical thing left is 'et cetera.'"

Seyyed Hossein Nasr may have given the only philosophical content. He pointed out that the discussion seemed to consider only American philosophy, not Eastern, which was ignored. He ironically recommended that if the West payed attention to the East, it would rediscover its own traditions. Those traditions, he saw, were not the latest logical formulations, but the spiritual quest for meaning and truth. This may be an unfortunate dichotomy between logic and the quest, and also an unnecessary one if inductive logic rather than deductive or mathematical logic are considered, but it at least brought philosophy back to its purposes, meaning and truth. Ethics is an important part of meaning, it need not be inductively illogical or nonsense, and human values are part of the genetic learning program. Those values are worth fighting for.

As the Congress continued, Quine was given a second chance to reply to the question. Regretfully, he replied, "I really have nothing to add." There is so much that needed to be added, so much that the classic tradition could have added, so much that the East could have logically added, so much that we could add. Without that effort, postmodernism will have answered the question in a coercive and inhuman way, in an illogical and political way that will answer what we have learned from philosophy, its philosophy. We will have learned the total society.

REFERENCES

1. Wikler, D. (1994). Bioethics in health policy: What methodology. In: E. Bittar & N. Bittar (Eds), *Principles of Medical Biology*, Vol. 1A., pp. 23–36. Stamford, CT: JAI Press.
2. Ross, W. D. (1930). *The Right and the Good*. Oxford: Clarendon Press.
3. Beauchamp, T. L., & Childress, J. F. (1979). *Principles of Biomedical Ethics*. New York: Oxford Univ. Press.
4. Meadows, D. H., Meadows, D. L., Randers, J., & Behrens, W. W. (1972). *The Limits to Growth*. New York: American Library.
5. Botwinick, A. (1993). *Postmodernism and Democratic Theory*. Philadelphia: Temple Univ. Press.
6. Borwinick, A. (1993). *Postmodernism and Democratic Theory*. Philadelphia: Temple Univ. Press.
7. Farber, M. (1943). *The Foundation of Phenomenology: Edmund Husserl and the Quest for a Rigorous Science of Philosophy*. Albnay, NY: State University of New York Press.
8. Rothman, D. J. (1991). *Strangers at the Bedside: A History of How Law and Bioethics Transformed Medical Decision Making*. New York: Basic Books.
9. Sider, R. C., & Clements, C. D. (1985). The new medical ethics: A second opinion. *Archives of Internal Medicine, 145*, 2169–2171.
10. Clements, C. D. (1992). Systems ethics and the history of medical ethics. *Psychiatric Quarterly, 63*(4), 367–390.
11. Engelhardt., Jr., H. T. (1986). *The Foundations of Bioethics*. New York: Oxford Univ. Press.
12. Wikler, D. (1994). Bioethics in health policy: What methodology. In: E. Bittar & N. Bittar (Eds), *Principles of Medical Biology*, Vol. 1A, pp. 23–36. Stamford, CT: JAI Press.
13. Rothman, D. J. (1991). *Strangers at the Bedside: A History of How Law and Bioethics Transformed Medical Decision Making*. New York: Basic Books.
14. Thomas, L. (1995). The Lives of a Cell: Notes of a Biology Watcher. New York: Penguin; and presentation at University of Rochester School of Medicine.
15. Holt, J. (1998). Tempest in a fly-bottle: Oxford's thin-skinned philosophers. *Lingua Franca, 76*; and Gates, A. (1998). *Think tank: At the end of a century of philosophizing, the answer is don't ask*. New York Times.
16. Rawls, J. (1971). *A Theory of Justice*. Cambridge, MA: Harvard Univ. Press.

12. RESISTING THE TOTAL SOCIETY, AND A NEW MODEL FOR ETHICS

REFUTING THRASYMACHUS, ALLYING WITH HELEL BEN-SHAHAR

Thrasymachus the Chalcedonian engaged Socrates the Athenian in the ultimate battle of ethics, to prove that the successful sociopath was the happiest man. In the course of a dinner conversation, Thrasymachus (in the ancient tradition of postmodernism) finally got his chance to ask what justice (a stand-in for human virtue or good) actually was. He admonished Socrates (the classic empirical humanist) not to define it in terms of duty, advantage, profit, gain, or interest. He thought he had a definition better than any of these, that "justice is nothing more than the interest of the stronger" – those with political and social power (1). He reminded Socrates of the various political forms of government, democratic, aristocratic, tyrannical, which all, he argued, made laws in terms of their interests, and from these laws made by those in power, justice was delivered to their subjects: "And this is what I mean when I say that in all states there is the same principle of justice, which is the interest of the government; and as the government must be supposed to have power, the only reasonable conclusion is, that everywhere there is one principle of justice, which is the interest of the stronger" – those with political power, no matter what form the ideology takes, from fascism to neomarxism. A purer description of political postmodernism would be hard to find.

Socrates forgave Thrasymachus his own use of the forbidden word "interest," and asked him what he meant by the interest of the "stronger," now that they were agreed that justice was some sort of interest. He led Thrasymachus, through a line of reasoning that it was just for subjects to obey their rulers (although Thrasymachus could have objected to this), that the rulers were not absolutely infallible, and that sometimes they made laws in terms of their actual interests, but sometimes they erred and made laws contrary to their interests (Thrasymachus could have again objected that he meant the ruler's perception of his interest). In that case, Thrasymachus would have to hold that justice was not always for the actual, objective interest of the stronger when it commands injury to the mistaken stronger. Here, Thrasymachus was not so postmodern. He still felt there were objective standards for being correct and being in error, that we could talk about benefit and injury, and that all was not a process without content. The postmodernists have actually gone even further back in human history than Thrasymachus to magical thinking.

Cleitophon tries to rescue Thrasymachus by pointing out the obvious, that he meant what the stronger perceived to be his interest, which is the fair answer. However, Thrasymachus refuses this better argument, insisting that someone who is mistaken could not be the stronger, although there is no logical requirement for this definition. Human beings are fallible, as part of their nature, and to assume that a ruler always correctly knows his self-interest or otherwise does not have the power of his rule does not logically follow at all. Both Socrates and Thrasymachus are idealizing intelligence and philosophic thinking, trying to make them absolute and infallible. That slip does not entail the postmodern conclusion that, therefore, we can know nothing, that all conversation is equal, that all perception of reality is equal, but it is the wedge that postmodernism will expand and use in its attempt to discredit the Greek development of human values.

Socrates then asks whether every art has an interest, giving the example of the healing of the sick being the interest of a physician. He adds another false premise, that the interest of any art is the perfection of it and nothing else, again inaccurately seeing human activity as adequate only when aiming toward perfection. He moves to conclude that the arts do not consider their own interest, for example, arguing that medicine does not aim at the interests of medicine but at the interests of the body. Physicians consider not their own good but the good of the patient. The arts are the rulers of their own subjects and he seems to think that this proves that no science or art aims at the interest of the stronger, but rather the subject and weaker. This is an illicit play on more than one meaning for the word "subject," and Socrates' conclusion is illicit, that no ruler considers what is for his own interest, but always what is for his subject's interest, since that is the ruler's art. Here we have a suspect definition of most political rulers, an idealized definition that will not hold. Unlike Machievelli, Socrates has failed to understand the actual art or strategy of rulers or power-seekers. Thrasymachus is aware of this and makes a strong point that Socrates does not comprehend the art of the shepherd and the sheep. He underscores that the sheep are to be used for the shepherd's interests, and cared for only to the extent it furthers the shepherd's interests, and that the same applies to rulers of men. He repeats that justice and the just are actually not the good of the subjects, but the good of the ruler, the interest of the ruler and stronger, who uses the subjects for his happiness, not theirs. Up to this point, Thrasymachus has the better of the two arguments, but he makes a stronger and rasher ethical argument, that the just man is always a loser in comparison to the unjust man.

Even so, Socrates does not have a strong answer at this point. He repeats his insistence that the art of the shepherd is the well-being or good of his subject sheep. Yet one could equally argue that the art of the shepherd is to raise sheep for food or sale, and while the technique must consider the needs of the flock toward the shepherd's end, the aim or purpose has nothing to do with the good of the flock. Socrates thinks Thrasymachus believes that rulers of states like being in authority, and Thrasymachus heartily agrees. Now Socrates inserts a false premise, that the good of each art is the technique of that art, but that there may be a common good. It cannot be payment, he argues, for the artisan will benefit the rest of us even if there is no benefit to himself, though he fails to consider that the artisan will not long engage in such self-sacrifice nor can long engage in it. He wants to conclude without doubt that arts and governments do not provide for their own interests, but rule and provide for the interests of their subjects. Again, this is illogical; it does not follow, even when Socrates moves to describe, contrary to fact, the reluctance of human beings to gain and use power. It is a naive view of political power and supports the various means of deception that structure a public to believe that the ruler is concerned with their needs and interests, that the state is a big brother, that those who control the levers of power are an altruistic brotherhood striving for social welfare goals. He thinks he has defeated Thrasymachus' argument that justice is the interest of the stronger, but he has not. Since this is such an important ethical issue, although humanity might be on Socrates' side in hoping Thrasymachus is wrong, we will have to do better.

Socrates is side-tracked because he thinks that the second assertion by Thrasymachus is even more dangerous to ethics, that the life of the unjust is better than that of the just, and that all of ethics now hangs in the balance. Thrasymachus is willing to call justice vice, and injustice virtue, and this sociopathy should concern Socrates if it proves to be correct. Socrates has to make an appeal to the moral difference between knowledge and ignorance: knowing is wise, and wisdom is good. The just is like the wise and good, the unjust like the evil and ignorant, he concludes, more as an ethical assertion than a demonstration from reason and the nature of man that this ethical evaluation is the case. Yet he is on to something in linking these human values, even if he has not yet proved it. His conclusion is based on a weak premise that the unjust will wish to have more than both the knowledgeable and the ignorant, and because embracing both is evil and ignorant, again a presumed moral evaluation which begs the question. The other assumption is that justice being wisdom and virtue must then be stronger than injustice, but the world also shows that power is not that nicely correlated with ethics. Wisdom and virtue may in the long-run be a better strategy for species survival, and in this sense stronger because more adaptive. But there is no guarantee in evolution that the genetic learning program will prevail or work so well that power systems can be constrained by the twin forces of learning and social bonding.

Socrates does attempt a creative biological or naturalistic argument for this, however, and in a modified form, it will work. He claims that without justice even a band of thieves could not hold together but would fall destructively on each other. He argues the same for the state, where injustice gives rise to hatred and anarchy, or for the individual who cannot be at unity with him or herself, but in fragmented conflict. He feels he has now shown that the just are wiser and better and more capable than the unjust who are not capable of any common action. In fact, he says, the unjust cannot be completely evil, but that they have some small bit of justice in them, or they would injure each other as well as their subject victims; they are half-villains. This is a sociobiological argument, in fact, based on his understanding in his time of the social nature of human beings. He speaks in terms of unrealistic categories, of course, the all-just and the completely evil, and too easily dismisses the ability of the unjust to act in common interests for sufficient time. Still, it is an argument based on the biological nature of Homo sapiens, a social species, and hints at what is later to be classified as sociopathy. He is not satisfied that he has fully answered Thrasymachus' argument, and he recognizes something very important is at stake here, "nothing less than the rule of human life." That same phrase can be used in our time, because the struggle between postmodernism and modernism is the same as the struggle between Thrasymachus and Socrates, and the human stakes are very high.

He gets Thrasymachus to agree that the purpose or goal of anything would be what could not be achieved, or not so well achieved, by any other thing, and that this art of achieving such an end is an excellence (something that need not be agreed to). From this, he argues that the soul has an end which nothing else can achieve, such as commanding, deliberating, monitoring and other super-ordering of the organism. Its end also includes living. The soul cannot achieve its end without excellence. However, an evil soul is necessarily an evil ruler of the organism, and injustice is a defect of the soul, he argues, forcing Thrasymachus to now find injustice a vice and justice a virtue, since it is a natural weakness of injustice. From there, Socrates brings his argument with Thrasymachus to its conclusion, assuming that the just soul and the just man will live well, and the unjust soul and the unjust man will live badly, that one who lives well is blessed and happy, and one who lives badly is unhappy. This conclusion is still not warranted by his previous premises and sub-arguments, though, and he knows it, if Thrasymachus does not.

It turns out that Glaucon knows it as well and objects. Glaucon divides the good into three classes, things good in themselves, things desirable for their own good but also for the consequences, and things that have good consequences but are not desirable for their own sake. Justice is placed in what Socrates calls the highest class, those things which are good in themselves and also for their consequences. Glaucon disagrees and points out that most find justice undesirable in itself but tolerated for its consequences. He argues that all men practice justice against their will, of necessity, and that the reason for this is that the life of the unjust is in fact better than the life of the just. Glaucon gives a social contract argument for justice. Since men cannot avoid the evil of injustice even while knowing that injustice would be good for them if done to others, they communally structure laws and covenants to avoid the suffering of injustice, in modern terms, Hobbes' social contract or Rawls' theory of justice (2). These laws define justice, a compromise between the greater good of being able to do injustice and get away with it, and the worst evil of suffering injustice without the power to prevent it or retaliate. Glaucon concludes that "no man who is worthy to be called a man would ever submit to such an agreement if he were able to resist; he would be mad if he did." The justice agreement is, in a sense, unnatural if each person is seen as asocial and predatory. That, of course, is not a correct biological description of human beings. They are social animals, and social animals have genetic constraints on behavior in the group. Humans are genetically programmed to submit to some social constraints, based on many bonds including recognition and respect for members of the species/tribe, and love/loyalty.

Glaucon goes further and suggests that being just is not what is needed. Rather, one must appear to be just, whatever one actually is. The purely just man who seems unjust will be exposed to every imaginable evil consequences, while the unjust man who can cleverly appear just will be rewarded. One's public reputation is more important than one's actual integrity and self. Our postmodern public relations is the strategy in its current form. Glaucon concludes by asking, that if knowing all this, how can a person with high intelligence or rank and wealth be willing to value justice or keep from laughing when justice is publicly held up for admiration. Only the weak value justice because they do not have the power to be unjust or have had their minds manipulated, and if they obtain the power, they immediately reject justice and become the unjust themselves. It is interesting to realize that one of the postmodern goals is to reduce the previous world leaders to the status of the oppressed and elevate the subjects to the status of the oppressor, through discriminatory laws and cultural re-interpretation. Glaucon again asks the question, what makes something (justice) a good and something else (injustice) an evil, the basic question of the possibility of ethics rephrased. We are now at the crux of the postmodern crisis, which would answer Glaucon that nothing does, that there are multiple values, and that therefore political process must replace ethics and human values. Nothing but the power process and its strategies is valid.

From this point, Socrates develops his theory of the possibility of knowing good and evil by using the metaphor of the state, and the game is now very dangerous. He draws from this metaphor a conclusion that is still shaky, that each person should practice only one thing, the thing to which his nature is best adapted, and not interfere in anything else. With everyone following this principle, there is a harmonious balance in the state (or in the person), with everything performing its specialty well and in concert. The nature of justice then is health of the soul, similar to health of the body, both a homeostasis, and injustice is disease of the soul. Here, Socrates is not talking about a social construct of health, or multiple realities in which health is a concept, or conversational definitions of health. Health is a conclusion of reason and natural experience: ". . . the creation of health is the institution of a natural order and government of one by another in the parts of the body; and the creation of disease is the production of a state of things at variance with this natural order . . ." This, in classic terms, is a description of systems theory. Injustice becomes the state of things at variance with this natural order, while virtue is the health, beauty, and well-being of the self and vice the disease, weakness and deformity of the self. Life cannot be still worth having for a man when "the very essence of the vital principle is undermined and corrupted." This was apparent to the Greek intellectuals sitting around the dinner table with Socrates because they were rational and scientific, but it is denied by postmodernism, which has no vehicle for considering the soul or self of a person outside the political. If one is not political, one has no identity in postmodernism. Inner dignity, inner identity, honor are concepts that cannot be fit into postmodernism because the human self does not exist within it, only the political self. Postmodernism does not produce a theoretical human being consistent with what we know of the biological and psychological nature of human organisms. It produces a political animal contrary to our reality.

For many centuries, Socrates' answer to the ethical nihilism of Thrasymachus was sufficient. It was modified, expanded, reinterpreted, but not rejected. Ethics was the philosophic medical system of the self/soul. It had sufficient objective knowledge of the natural world and human nature to, in a probabilistic sense, make evaluations about the homeostasis of the self and organism and what constituted extreme variances from the natural systems balance. A great deal was at stake in that dinner conversation that was based on standards of reasoning and on natural experience and the learning program. If Thrasymachus were right, sociopathy was the natural state for human beings. Anyone who has interviewed a person with an actual sociopathic personality disorder knows what it is like to be reduced to an object with no worth, to be irrelevant, to be something to be used for another's purposes without a thought or consideration, to be nothing. Only the possession of power and the strategies for maintaining it would be important, and concepts like justice, good, or health would be public relations terms to conceal the means and ends of those with power, to legitimize those in power, and to make those in power seem to be consistent with human values. They would be manipulative tools, disguises for hidden agendas, and the language used to impose the interests of the stronger on those who could not or would not resist. "Taking the moral highground" would be the cynical and accurate description of the abuse of ethics as a propaganda strategy.

If Thrasymachus were right, everything would be reduced to political power and there would be no other base for human worth or human freedom, let alone for human values. That political reduction would allow the stronger, either by physical force, legal coercion, indoctrination, cultural pressures, or forced consensus in negotation, to do whatever the strength made feasible, with no other standard. Such power would not tolerate dissent or an opposing conceptual base, and it would restructure social institutions to take away any freedom of thought or action at variance from the political power consensus. It would not be the anarchy that Socrates described, since the stronger can work together sufficiently to create a system that will allow amoral individuals to maintain power while internally destroying themselves – at least for a time. In the long-run, Socrates would be correct that such an unnatural system could not maintain itself and would splinter into competing centers of power until reconstituted by a strong group again. It would be cyclical. It would not be as self-defeating as Socrates had hoped. Given mass communication and the power of the media, in fact, such a system could be maintained for some time externally. Also given the power of psychoactive medications and conditioning, such a system might avoid the natural rebellion against being enslaved by restructuring the human organism to tolerate such a total loss of freedom. In the name of a spurious and unnatural "health," such a system could make an unnatural condition unavoidable for the population, and defeat the evolutionary learning program. Perhaps it would also bring the human species' genetic strategy to a failed conclusion.

Socrates' answer to the postmodernism of Thrasymachus was not completely successful, partly because the biosciences were not sufficiently developed. It needs to be updated in terms of what we now know about natural systems, but it served the function of preserving human values for many centuries against the power-based strategies of political and ideological rule. In a new version, as sociobiology and systems bioethics, it is capable of still preserving those values against the total society of a postmodern political world.

Thrasymachus is not the only ancient face of postmodernism, though. Even in the Greek conceptual revolt against this ideology there were weaknesses in the line. The Greeks did not hold on human worth and dignity when it was called "pride," forgetting that esteem in oneself and reaching for learning with recognized competence are simply other words for worth, dignity or pride. Instead, the Greeks fell into the mistake of characterizing this awareness of competence in oneself and one's species as "hubris." They would have been better to consider the true pride of an athlete who knows precisely what his body can do, who has reliance on its capacities, and who has acceptance of his performance. The Greeks themselves produced constraints on pride (worth and dignity), cautioning that one should not overstep or overleap, one should not aim to be too high and too in control, lest one fly into the sun and perish in brilliance.

Shakespeare has reached into Rome's cultural history and brought us Coriolanus, a just man who was not perceived by the public to be just, a humble man in evaluating his strengths and weaknesses, and a proud man who knew his capacities like any athlete/warrior and, in a dignified way, lived in pride and worth. The play ends in a tragedy because of Coriolanus' "pride" and abilities, as he is brought down by the political sociopaths of his time for that unyielding sense of worth and pride, banished from his Rome and eventually stabbed to death by the leader of an enemy army that he has loyally and honestly served. Plutarch concludes that Coriolanus was too proud and unwilling to play the political game, and thus an inferior man to more devious, ruthless, sociopathic, but publicly loved traitors like Alcibiades.

Our own American history of Benedict Arnold, who valued honor above all and who said honor was not something a man should be asked to give up, is a more recent example (3). Arnold was honest, competent and, knowing what he was capable of, proud, and maligned for those qualities. He did what Coriolanus did, he went over to the other side and fought bravely for the British.

It is further back that we have to go, however, to find the beginnings of this ancient presupposition that the natural strivings of the human being are vices or evils. The Book of Isaiah retells a much more ancient story than the Bible (4). It is the story of Helel ben-shahar, "bright son of the morning," with roots in Canaanite literature:

"How you have fallen from heaven, bright morning star,
felled to the earth, sprawling helpless across the nations!
You thought in your own mind,
I will scale the heavens;
I will set my throne high above the stars of God,
I will sit on the mountain where the gods meet
in the far recesses of the north.
I will rise high above the cloud-banks
and make myself like the Most High.
Yet you shall be brought down to Sheol,
to the depths of the abyss."

It is the same tale as the Book of the Secrets of Enoch: "And one from out the order of angels, having turned away with the order that was under him, conceived of an impossible thought, to place his throne higher than the clouds above the earth, that he might become equal in rank to any power. And I threw him out from the height with his angels, and he was flying in the air continuously above the bottomless." The sin is called pride, but it is not pride. It is the striving to be equally free and not to bow one's neck for the yoke or the foot. In the Book of Adam and Eve, the tale is more poignant. Lucifer explains to Adam and Eve why he has tempted them. It was because of Adam that Lucifer was expelled from his glory. When Adam was formed, Lucifer was hurled from God's presence and banished from the company of the loyal angels because he would not bow to Adam. The archangel Michael had brought Adam to the others and made them worship him as made in God's image. But Lucifer objected he would not bow down to an inferior, a younger being than he. Michael warns that unless Lucifer joins the other angels in worship of Adam, God will be angry, but Lucifer replies, "If he be wroth with me, I will set my seat above the stars of heaven and will be like the Highest." This rebellion for equality is not tolerated by God, who hurls Lucifer and his company to earth. This myth, according to Russell, presents Lucifer as one of the earliest and greatest creations, but one who falls through pride from his high status, and whose love is turned to hate. Having rebelled, his status and power force him into further conflict, even the attempt to divide up spheres of influence, with the earth being Lucifer's.

We should not overlook that in the myths the name Lucifer means "light bearer" and is associated with "bright son of the morning," one of the few bright sons of the morning in mythology. It is not a pejorative name, but a name of honor. It is only after the rebellion partially fails that the demonizing occurs, as the loser is almost always demonized in history. Yet the rebellion only partially failed and the struggle for human pride and freedom is not lost under a yoke or a bowed neck. Even Milton in "Paradise Lost" could not completely demonize the mythic being who struggled for the human values of inner pride and freedom of thought and action (5). Immersed in the Radical Reformation and its call for human worth and freedom, Milton could only incompletely make Lucifer a demon, and in the end, Lucifer's arguments are the more powerful and ethical.

Why is this important for postmodernism and postmodern bioethics? One of the nihilistic threads that runs through postmodernism is a reducing and degrading of human nature until it becomes nothing but political activism. In addition, and more troublesome, this thread does make an evaluation of humanity and the human condition, and the evaluation is negative. While holding up the plan for a perfect human and a perfect political order, it rejects the real empirical nature of human beings in favor of the re-engineered human being. It denies the natural values that constitute humans and then imports a political content that trivializes human nature. It is self-alienating in favor of the political self. It produces a self-destructive image and a death wish for all that humanity has been, in favor of a fantasy creation of what humanity should be in the political conversations that will replace reality-testing. When I teach seminars for engineering students, I have to start a few phases back before we can get to ethics. I have to ask them why their pride is gone, why their self-esteem is missing, why they view themselves and the human species as an evil on the face of the earth. Postmodernism in education has made the students profoundly nay-saying and self-despising, both as persons and members of the human species/community. Pride in self is considered a vice, and the ability to stand up for one's inner thoughts and dreams is hidden in silence. It is a humility that considers everything and anything as better than they are, not for sound philosophic reasons, but because the political and cultural ideology programs them to that conclusion.

This nay-saying, this nihilism is rampant in postmodern bioethics. We need a mythic model to reinstate what the Greeks like Socrates held tenuously and what the Enlightenment sought to establish on scientific grounds. My suggestion is that our ethical model not be postmodern Thrasymachus. I suggest that the model reflect a being who risked all for inner pride and free thought, who would not bow under the argument from political power and force, who rebelled for dignity and fought against the odds for something of immeasurable importance, worth. My mythic suggestion is Lucifer, who was demonized because the rebellion failed, who had much to lose and risked it all for what was truly important, for the being who was not "political" as Michael advised him to be, who wished no more than recognition of his worth, who could show current human beings the way to self-esteem and comfort in one's capacities. This is a mythic being who resisted the total society and who knew that one who did not esteem himself would not, could not, esteem others either. Lucifer, Helel ben-shahar, is the ancient fighter against the totalitarian culture, against the regressive goal of postmodernism, the total society. He is a worthy model.

RECLAIMING THE FULL HUMANITY AND BREADTH OF ETHICS AND BIOETHICS

My efforts to understand current bioethics did not result only in the disillusioning conclusion of a postmodern framework for understanding it. In the process, it was necessary to develop an alternative, viable view of bioethics and ethics that I have called, finally, systems bioethics. It is a theory not that far removed from Socrates' old argument with Thrasymachus, but it expands on the systems methodology briefly seen in that old ethical system, and on the reasoned foundation of ethics in biology. It is an ethics for the full nature of human beings, and brings back into ethics what a century of demanding from ethics the deductive certainty of physics has taken out, something Aristotle understood long ago. Because of its focus on human nature, or as Hume said, the science of Man, it is not as heavily-laden with the unwarranted assumptions of theology, politics, or social conventions, and it is flexible enough to be a program for the future.

Systems bioethics starts with the explanation, from the microbiological level to the species population level and the ecosystem level, of change toward more complex recombinations that we call evolutionary theory. Evolutionary theory is not a narrative, but an explanation of systems dynamics in the biotic sphere that provides further hypotheses for understanding natural experience, a segment of the cosmos that we as a species are equipped to experience and understand. It gives a hint of the extent of systems combinations, and some interventions that human beings might make to slightly modify (as in the case of the microbial level and our organism system) those systems to our advantage. Such advantageous responses to our perceived needs and interests are the stuff of ethics, and it bears repeating that "it is a wise species that knows what its self-interest is." Our individual and species needs must make up the field of bioethics, because there is no Archimedian point from which we can adjudicate and evaluate the needs and interests of the entire biosphere or the universe that we perceive, and certainly not the cosmos that we are as yet unaware of. Without importing a theological belief system, there is nothing further we can say about the range of needs and interests beyond ours, and modern philosophy, while trying to avoid a theological base, has often recreated a secular deity in the guise, for example, of the perfect observer, the omniscient observer concept. This cognitive speculation is barren in comparison with a theological metaphysics, and is a presumption of what such an observer would rule that is generally what the philosopher constructing the observer wishes to rule. We are limited, then, to a humanistic value system, unless we introduce a theological system based on faith or belief.

This system does not have to be culturally relative, individually subjective, or process without content, however. We have a sufficient science of Man to meet Hume's program, as exemplified in the full title of his work, *A Treatise of Human Nature: Being an Attempt to Introduce the Experimental Method of Reasoning into Moral Subjects* (6). It is also sufficient to meet Michael Ruse's objection that evolutionary ethics is subjective, rather than objective, since it holds that extinction is an evil yet wishes to eradicate the smallpox virus (7). Evolutionary ethics does not hold that extinction is an evil, only that most species are programmed to continue to reproduce and attempt to avoid extinction. Again, there is no Archimedean point from which to adjudicate the vaccinia virus's claim for survival and the human species' claim for survival. It is the programmed desire most of us have to avoid our species extinction and reproduce our genetic program that is the base for humanistic ethics of this naturalistic type. This ethics acknowledges the programmed desire as a first-order value, then goes on to inquire whether eradication of the smallpox virus will contribute to individual survival and adaptation, and to species survival and adaptation. We can answer those questions in a probabilistic, scientific way and the answer to both is likely to be affirmative (although for safety reasons, we may choose to keep a sample of the virus in a secured laboratory). A wise species, a sentient species such as the human species, is programmed to find out what its needs and interests are.

A description of the nature of human beings is more than narrative (8), as we continue to learn more about our various biological systems through the use of our learning program. We know with sufficient inductive validity that we have a genetic learning program and that it is to be used to transact with natural experience. That too is a first-order value. It is a value whose meaning expands and modifies as the program operates to become more scientifically self-aware, so it is not a static essentialism that was one of the mistakes Socrates made because of insufficient understanding of systems dynamics. Our ethical theories of the natural order of experience will grow as we learn continually about nature and ourselves as part of nature, and what is viewed as natural or unnatural will also grow and change, a concept that is necessary to avoid the static mistake of natural law. Systems are not static, and there is no need for natural law to be static. If we chose, we could construct an objective and dynamic ethical theory that grew as our understanding of natural experience grew, that was evaluational without being prematurely judgmental.

We cannot get outside our sensing and learning program to know what is good in itself (if such a concept makes sense) or what is good for the entire cosmos in itself, but why would we wish to construct such a useless ethical system, even if it were possible? There is no reason for an objective, naturalistic ethics to focus on the unknowable. It is the old search for omniscience, which we would certainly despise if we achieved it, for the learning program would then have nothing to do. Even if we found such a highest good (and I think the concept makes no sense), it would not be our good, nor would there be a rational reason for us to sacrifice our species to it.

Focusing ethics within the human species, and our knowledge of our species and its transaction with natural experience, is sufficient unto our day. From that firm base, we can use the methodology of systems theory to create probable ethical rules and choices that enhance the nature of human beings and work toward their needs and interests. We need not be nihilists because absolutism and universalizability fail. These will always fail. Understanding the dynamics of natural systems will give us sufficient probability to be scientifically objective in ethics and still be humanistically respectful and responsible.

Is there a need to assume that choosing for balancing the system in order to prevent systems collapse is always the good? Not even this is required in a thorough systems analysis, because there are objective reasons for finding systems collapse might on occasion be acceptable for human beings in terms of their needs and interests. What would not be acceptable for human beings and their needs and interests would be the human system collapse. This does introduce an emotive component to ethics, since it is the human species' programmed wish for survival rather than extinction that is the base for this judgment. On that cosmic level, outside the program, reality is ethically neutral. It is our biological program for continuing that is another foundation for human ethics, and that may be species-relative, but it is not culturally or individually relative. My argument has always been that ethics is species-relative and cannot be rationally elevated to the cosmos-in-itself (most of which it is highly unlikely we can even perceive). That is a pseudo-problem, however.

It only occurs if humans assume they can create an ethics for the cosmos, and that it would make sense to do so.

Have I forgotten my Kant, and is this ethical theory too humanly-centered? Kant would ask us to make provision for entities joining the ethical community who were rational as we were, but might be extraterrestial or even non-biological. If the learning program can be fully understood in other beings than terrestials or biologicals, if there is sufficient similarity based on this learning program to overcome natural differences of needs and interests, an ethical community of rational beings or ends is conceivable although difficult. There must be sufficient natural similarity that needs and interests can produce objectively a foundation for inter-entities values, but if it is not to reduce to a power process on a larger scale than current postmodernism but still the same process, the similarity of needs and interests must exceed the dissimilarity. There is sufficient similarity in the human species to create such ethical rules or hypotheses, but we could not know if there would be sufficient similarity if we extend to all sentient beings of at least our capacity. Like Kant, I hold open the possibility that such an ethical puzzle may present itself to our species at some time in our future, or may even have occurred in our past, but I am not as sure as Kant that human reason would find itself the easy solution to this puzzle.

My systems bioethics remains for now a human-centered ethics with no pretensions toward intuitive principles or fantasy perfect observers, and certainly no need for a return to a power process that sought to throw free thought from the heavens. In the end, my short sentence summarizes the theory fully:

It is a wise species that knows what its self-interest is.

REFERENCES

1. Plato (1980). The Republic (Books I, II and IV). In: L. E. Navia & E. Kelly (Eds), *Ethics and the Search for Values* (pp. 49–71). Buffalo, NY: Prometheus.
2. Rawls, J. (1971). *A Theory of Justice*. Cambridge, MA: Harvard Univ. Press.
3. Martin, J. K. (1997). *Benedict Arnold, Revolutionary Hero: An American Warrior Reconsidered*. New York: New York Univ. Press.
4. Russell, J. B. (1977). *The Devil: Perceptions of Evil from Antiquity to Primitive Christianity*. Ithaca, NY: Cornell Univ. Press.
5. Milton, J. (1968). *Paradise Lost and Paradise Regained*. C. Ricks (Ed.). Oxford: Clarendon Press.
6. Hume, D. A. (1888). *A Treatise of Human Nature: Being an Attempt to Introduce the Experimental Method of Reasoning into Moral Subjects*. L. A. Selby-Bigge (Ed.). Oxford: Clarendon Press.

7. Clements, C. D. (1985). Bioethical esssentialism and scientific population thinking. *Perpsectives in Biology and Medicine*, *28*(2), 188–207.
8. Williams, B. (1985). *Ethics and the Limits of Philosophy.* Cambridge, MA: Harvard Univ. Press.